Embryology:
Review for New National Boards

Embryology:
Review for New National Boards

Raymond F. Gasser, Ph. D.
Professor of Anatomy
Louisiana State University Medical Center
New Orleans, LA

Kurt E. Johnson, Ph. D.
Professor of Anatomy and Cell Biology
George Washington University Medical Center
Washington, D.C.

Ronald C. Bohn, Ph. D.
Associate Professor of Anatomy and Cell Biology
George Washington University Medical Center
Washington, D.C.

J&S

J&S Publishing Company Inc., Alexandria, Virginia

J&S

Composition and Layout: Ronald C. Bohn, Ph. D.
Cover Design: Kurt E. Johnson, Ph. D.
Printing Supervisor: Robert Perotti, Jr.
Printing: Goodway Graphics, Springfield, Virginia

Library of Congress Catalog Card Number 97-070353

ISBN 1-888308-01-X

Raymond F. Gasser would like to dedicate his portion of this book to Rusty, his loving wife for 36 years. Kurt E. Johnson would like to dedicate his portion of this book to my wife Julie M. Okkema, M.D. and my children Melissa, Abraham, Justine, and Alexander. Ronald C. Bohn would like to dedicate his portion of this book to Susan and Eric. Finally, J & S Publishing Company, Inc. would like to dedicate this entire book to the memory of Alvin Mercer, a recently departed employee. Your work and friendship were highly valued. Rest in peace, dear friend.

Table of Contents

Preface

This book is designed to enable you to review in just 1-2 days all of the basic embryology you studied in the first year of medical school. We have been able to condense a review of all basic embryology into a single book because of the new format of the National Board, Step 1 Exam. It is no longer prudent to review exhaustively the basic science courses because the new examination format no longer rewards an encyclopedic knowledge of the basic sciences. Instead, the new exams test knowledge of the scientific basis of disease and the ability to apply basic scientific information to the clinical reasoning process. Consequently, the most efficient way to study for the new exam is 1) to review only the most clinically relevant material from each basic science course and 2) to focus on the application of this material to the solution of clinical problems. These two new study features form the core of this text.

If you answer every question and read all the tutorials in this book, you can cover within 2 days all of the most clinically relevant information from your basic embryology courses. You will find that many embryologic facts reviewed or learned anew will be presented in the context of a clinical case or an illustration. We hope that the clinical cases and illustrations will enhance your understanding and recall of the information. Finally, you will learn from the tutorials how embryologic information is used by knowledgeable physicians to understand the courses of diseases and the significance of abnormal findings.

Raymond F. Gasser, Ph. D.
New Orleans, LA
Kurt E. Johnson, Ph. D.
Ronald C. Bohn, Ph. D.
Washington, D.C.
April, 1997

vii

Acknowledgements

The authors would like to thank Rowena Spencer, M.D., Eugene Wolf, Robert Musselman, D.D.S., and Theodore Thurman, M.D. for the use of their photographs of malformations. The authors are grateful for all of this help and acknowledge any errors in this book as their own.

Disclaimer

The clinical information presented in this book is accurate for the purposes of review for licensure examinations but in no way should be used to treat patients or substituted for modern clinical training. Proper diagnosis and treatment of patients requires comprehensive evaluation of all symptoms, careful monitoring for adverse responses to treatment and assessment of the long-term consequences of therapeutic intervention.

Figure Credits

Figures 1.3 and 10.5 From Kramer, R.L., Obstetrics and Gynecology: Review for New National Boards, ©1996 reprinted with permission of J & S Publishing, Alexandria.

Figures 2.2, 3.1, 3.5, 3.7, and 7.1 From Abeloff, D., Medical Art: Graphics for Use ©1982 reprinted with permission of Williams and Wilkins, Baltimore.

Figure 3.3 From Taeusch, H.W., Ballard, R.A., Avery, M.E. (eds), Schaffer and Avery's Diseases of the Newborn, ed 6, ©1991 reprinted with permission of W.B. Saunders, Philadelphia.

Figures 3.4, 5.6, 9.3, 10.1, 10.2, 10.3, and 10.4 From Johnson, K.E. and Slaby F.J. Anatomy: Review for New National Boards, ©1992 reprinted with permission of J & S Publishing, Alexandria.

Figures 5.3, 5.4, and 5.7 ©1981 reprinted with permission of John Wiley & Sons, New York.

Figures 6.2 and 9.4 From Geelhoed, G.W., Surgery: Review for New National Boards, ©1995 reprinted with permission of J & S Publishing, Alexandria.

Figure 8.3 and 8.4 ©1972 reprinted with permission of Lippincott-Raven, Philadelphia.

Figure 9.5 From Lentz, T.L., Cell Fine Structure, ©1971 reprinted with permission of W.B. Saunders, Philadelphia.

CHAPTER I
GAMETOGENESIS AND EARLY DEVELOPMENT

<u>**Items 1-10**</u>

Examine the photomicrograph in **Figure 1.1** below and then match the labeled structure with the **MOST** appropriate description of its functional role in gametogenesis in the items that follow. Answers may be used once, more than once, or not at all.

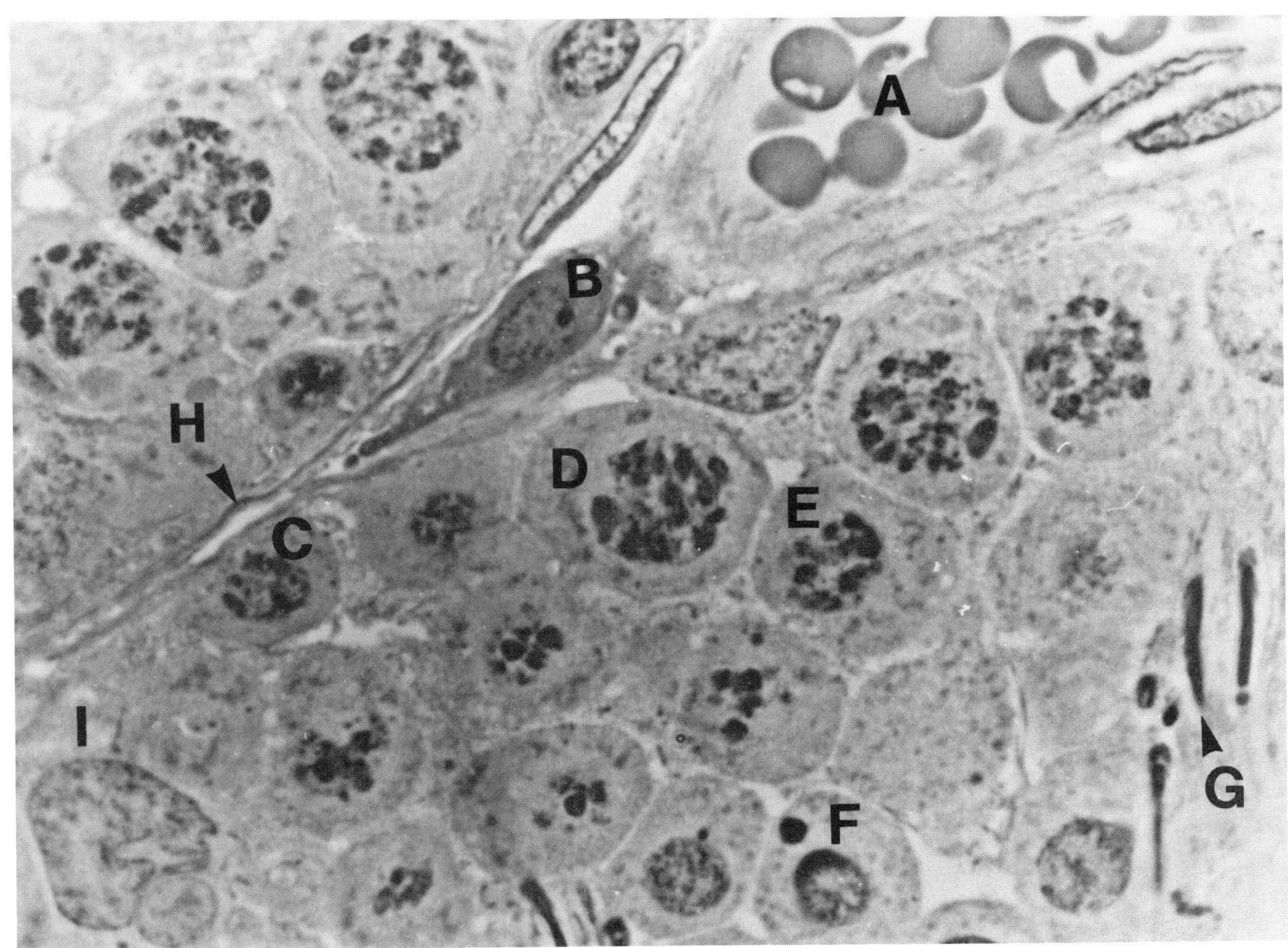

Figure 1.1

1.   This structure conveys endocrine secretions of the testis into the systemic circulation.

2.   This structure secretes an androgenic hormone required for spermatogenesis.

3.   This structure releases androgen binding protein which is required for spermatogenesis.

4.   This structure is a diploid proliferative cell.

5.   This structure contains paired bivalents and it is the site of formation of chiasmata and crossing over.

6.   This structure has just completed the first meiotic division.

7.   This cell is haploid and is in the early stages of spermiogenesis.

8.   This cell is haploid and is in the late stages of spermiogenesis.

9.   This cell is the first generation descendant of the primordial germ cell.

10.   This cell is the only descendant of the coelomic epithelium in this photomicrograph.

ANSWERS AND TUTORIAL ON ITEMS 1-10

The answers are: **1-A; 2-B; 3-H; 4-C; 5-D; 6-E; 7-F; 8-G; 9-C; 10-I**. **Figure 1.1** shows a high power photomicrograph of a portion of two **seminiferous tubules** in the human adult male. The testis develops from **primordial germ cells** (precursors of gametogenic cell line), coelomic epithelium over urogenital ridge [precursor of **Sertoli cells** (I)], and urogenital ridge mesenchyme (precursor of the tunica albuginea, interstitial vascular and connective tissue, and Leydig cells).

The interstitial tissue between these seminiferous tubules contains a single endocrine **Leydig cell** (B) which is an important source of androgenic steroids such as **testosterone**. The testosterone is conveyed to the systemic circulation via capillaries (A) and also diffuses to become concentrated by the action of **androgen binding protein**, which is synthesized in **Sertoli cells** (I) in the seminiferous epithelium and is essential for maintaining the high local androgen concentrations required to support spermatogenesis.

Spermatogenesis begins in diploid (2N DNA, 2N chromatids in 2N chromosomes) proliferative cells called **spermatogonia** (C) which rest on the basement membrane (H) of the seminiferous epithelium. Spermatogonia and all of their descendants are derived from primordial germ cells. The spermatogonia are a stem cell population where one daughter remains as a proliferative spermatogonium and the other differentiates into a **primary**

spermatocyte (D). Following a round of DNA synthesis and chromatid duplication without centromere division, the primary spermatocyte (4N DNA, 4N chromatids in 2N chromosomes) becomes a tetraploid cell with homologous chromosomes pairing to allow **genetic exchange** between sister chromatids during chiasmata formation. After crossing over, the first meiotic division occurs, producing **secondary spermatocytes** (E) with a haploid number of double chromosomes (2 N DNA, 2 N chromatids in 1N chromosomes). In the second meiotic division, centromeres divide without either DNA synthesis or chromatid duplication. Once the second meiotic division occurs, **spermatids** (F) are produced which contain 1N DNA, 1N chromatids in 1 N chromosomes. A complex cytomorphogenetic process of **spermiogenesis** then converts the spermatids into **spermatozoa** (G).

Items 11-15

Match the spermatid organelle in the answers with the **MOST** appropriate description of the changes that occur in it during spermiogenesis. Answers may be used once, more than once, or not at all.

(A)	Golgi apparatus
(B)	Nucleus
(C)	Mitochondria
(D)	Centrioles
(E)	Plasma membrane
(F)	Peroxisomes
(G)	Rough endoplasmic reticulum
(H)	Smooth endoplasmic reticulum
(I)	Crystalloid inclusion bodies

11. Forms the single, locomotory flagellum of the spermatozoon.

12. Becomes completely euchromatinized, condensed, and streamlined.

13. Forms an elaborate spiralling sheath in the middle piece of the spermatozoon.

14. Forms a structure with an abundance of tubulin and dynein.

15. Forms the acrosome.

ANSWERS AND TUTORIAL ON ITEMS 11-15

The answers are: **11-D; 12-B; 13-C; 14-D; 15-A**. **Spermiogenesis** is the complex cytomorphogenetic process whereby the haploid spermatids are converted into the spermatozoa, specialized cells capable of fertilizing an ovum. During spermiogenesis, various spermatid organelles make significant contributions to the developing spermatozoon. For example, the **Golgi apparatus** (A) forms the acrosome, a modified lysosome which aids in penetration of the corona radiata and zona pellucida surrounding the ovum. The **nucleus** (B) becomes highly condensed, euchromatinized (and therefore nonfunctional), and streamlined. **Mitochondria** (C) migrate to the pole of the spermatid opposite the acrosome and form a spiralling sheath of energy ATP-producing organelles around the flagellum in the middle piece of the spermatozoon. Simultaneously, the pair of **centrioles** (D) migrate to the same pole of the spermatid and one of the pair elaborates a flagellum, the locomotory organelle of the mature spermatozoon. Flagella have many microtubules and an abundance of dynein. The entire spermatid is surrounded by a **plasma membrane** (E). During fertilization, the plasma membrane of the one fertilizing spermatozoon fuses with the plasma membrane of the ovum and the haploid male pronucleus enters the cytoplasm of the ovum. Next, the male and female pronuclei fuse during syngamy to form the diploid zygote nucleus.

Spermatids have an abundance of **rough endoplasmic reticulum** (G) which is important for synthesis of proteins that will become packed into the developing acrosome by the Golgi apparatus. **Peroxisomes** (F), **smooth endoplasmic reticulum** (H), and **crystalloid inclusion bodies** (I) are not important features of spermatids and play little crucial role in spermiogenesis.

Examine the photomicrograph in **Figure 1.2** below and then match the labeled structure with the **MOST** appropriate description of its functional role in development in the items below.

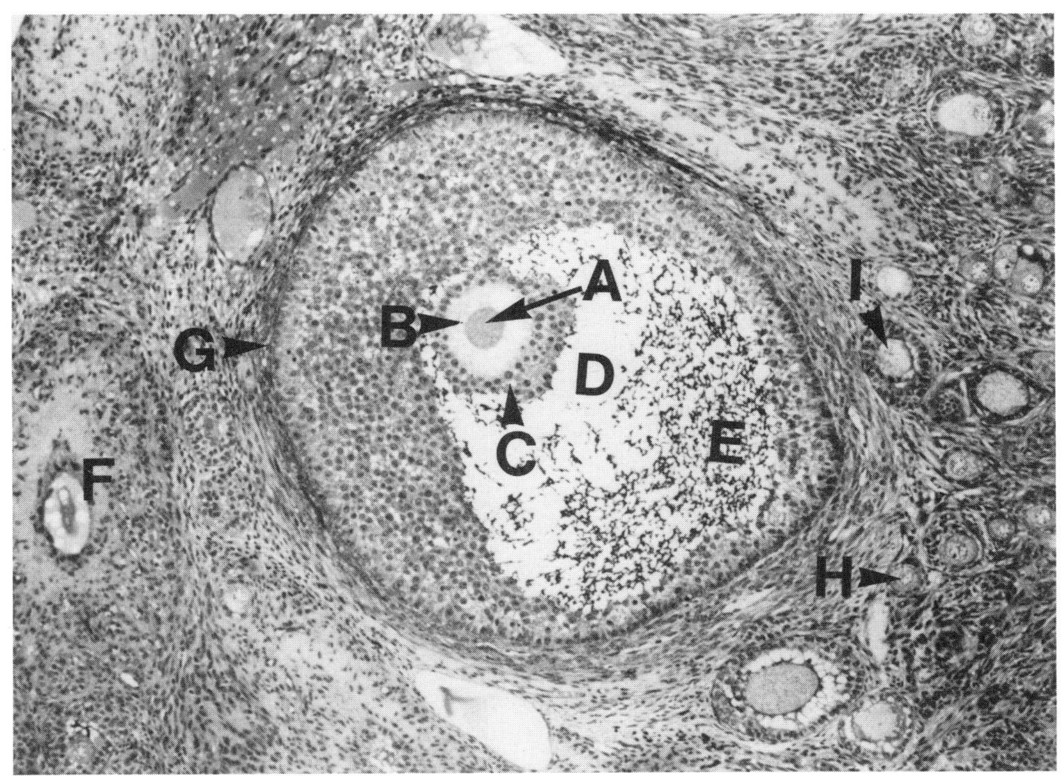

Figure 1.2

16.   This structure will ultimately become a haploid ovum.

17.   This structure is shed along with the ovum and will become the corona radiata.

18.   This structure is a primordial follicle.

19.   This structure is a viscous secretion product of the granulosa cells called the liquor folliculi.

20.   This structure will become theca lutein cells after ovulation.

ANSWERS AND TUTORIAL ON ITEMS 16-20

The answers are: **16-A; 17-C; 18-H; 19-E; 20-G**. **Figure 1.2** is a photomicrograph of a histological section of a primate ovary. A **primordial follicle** (H) is an ovum surrounded by a simple squamous epithelium of granulosa cells. **Primary follicles** (I) have multiple layers of granulosa cells but no follicular antrum (D). Secondary or antral follicles have many layers of granulosa cells and an antrum filled with a viscous glycoprotein secretion of granulosa cells called the **liquor folliculi** (E). An antral follicle contains a **primary oocyte** (A) surrounded by a **zona pellucida** (B) and a layer of granulosa cells, the **cumulus oophorus** (C). This layer of granulosa cells will be shed at ovulation along with the oocyte and remains attached to the zona pellucida thereby forming the **corona radiata**. External to the basement membrane of the follicle, is located the **theca interna** (G) that is composed of steroid-secreting cells that will be converted into theca lutein cells of the corpus luteum after ovulation. An **atretic follicle** (F) is also labeled.

Examine the scanning confocal micrograph in **Figure 1.3** below and then choose the **MOST** appropriate answer to the items below.

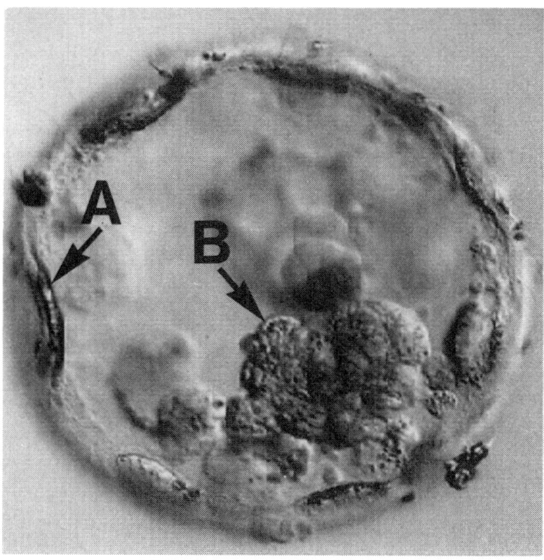

Figure 1.3

21.   All of the following are true for this developmental stage **EXCEPT**:

(A)   It has hatched from the zona pellucida.
(B)   It is less than 24 hr post-fertilization.
(C)   It has a well defined trophoblast.
(D)   It has a well defined inner cell mass.
(E)   It is more advanced than the stage typically used for embryo transfer during *in vitro* fertilization.

22.   Which of the following structures is normally derived from the structure labeled **A**?

(A)   cytotrophoblast
(B)   fetal red blood cells
(C)   embryonic ectoderm
(D)   lining of gastrointestinal tract
(E)   epiblast

23.   Which significant developmental process will occur next after the stage illustrated
        here?

        (A)   compaction
        (B)   blastocyst formation
        (C)   blastocyst hatching
        (D)   implantation
        (E)   gastrulation

24.   Which statement **BEST** describes the developmental fate of cells in the structure
        labeled **B**?

        (A)   They form predominantly yolk sac.
        (B)   They form only skin and derivatives.
        (C)   They contribute to the central nervous system.
        (D)   They are precursors of the placenta.
        (E)   They form predominantly fetal RBCs.

25.   In a four cell cleaving zygote, an individual blastomere is a precursor of:

        (A)   **A** derivatives only
        (B)   **B** derivatives only
        (C)   Both **A** and **B** derivatives
        (D)   Neither **A** nor **B** derivatives

ANSWERS AND TUTORIAL ON ITEMS 21-25

The answers are: **21-B; 22-A; 23-D; 24-C; 25-C**. **Figure 1.3** is a scanning confocal
micrograph of a human **blastocyst**. It is derived from an ovum fertilized *in vitro*. At this
stage, the embryo is 4-5 days post-fertilization. It has shed the **zona pellucida** and is ready
to begin implantation.

It has differentiated into two major structures, a **trophoblast** (A) and an **inner cell
mass** (ICM) or embryoblast (B). Recent experimental studies using human 4-cell cleaving
zygotes derived from ova fertilized *in vitro*, show that each blastomere can make a
contribution to both the trophoblast and the inner cell mass. The trophoblast is destined to
form the cytotrophoblast and syncytiotrophoblast, major contributors to the developing
placenta. After implantation the ICM forms two layers, the **epiblast** and **hypoblast** and is
called the bilaminar disc stage. Next, during gastrulation, the bilaminar disc forms three
primary germ layers. The ICM forms all three primary germ layers and thus forms all
ectodermal, mesodermal, and endodermal derivatives in the embryo.

Match the kind of gametogenic cell in the answer with the **MOST** appropriate functional description of the gametogenic cell listed in the items below. Answers may be used once, more than once, or not at all.

    (A)   Primordial germ cell
    (B)   Primary oocyte
    (C)   Primary spermatocyte
    (D)   Secondary oocyte
    (E)   Secondary spermatocyte

26.   A tetraploid cell produced by the millions after puberty.

27.   A cell with 2N DNA that undergoes an equal cytoplasmic division to form spermatids.

28.   A cell with 2N DNA that undergoes unequal cytoplasmic division.

29.   A cell derived from the yolk sac that forms spermatogonia.

30.   A cell with 4N DNA that can be arrested in meiosis I for 40 years or more.

31.   All of the following statements concerning gametogenesis are true **EXCEPT**:

    (A)   More male gametes are produced than female gametes.
    (B)   Meiosis in males is completed in the testis.
    (C)   Meiosis in females is completed just before ovulation.
    (D)   The secondary oocyte has 2N DNA.
    (E)   The second polar body is a haploid cell.

32.   All of the following statements concerning spermatogenesis are true **EXCEPT**:

    (A)   It ends around 55 years of age.
    (B)   Spermatogonia are present prior to puberty.
    (C)   Spermatozoa are present after puberty.
    (D)   Primary spermatocytes have paired homologs in them.
    (E)   Spermatids are haploid cells with 1N DNA.

ANSWERS AND TUTORIAL ON ITEMS 26-32

The answers are: **26-C; 27-E; 28-D; 29-A; 30-B; 31-C; 32-A**. The **primordial germ cells** (A) are diploid proliferative cells derived from the wall of the yolk sac. They differentiate into diploid **spermatogonia** and **oogonia**. **Primary spermatocytes** (C) are tetraploid cells that are produced in vast quantities by mitosis of spermatogonia in the male after puberty. The proliferation of spermatogonia and production of primary spermatocytes continues until death. **Primary oocytes** (B) are tetraploid cells which are arrested in meiosis I from their time of formation (before birth) until ovulation at some time later, up to the menopause. **Secondary oocytes** (D) are diploid cells that undergo unequal cytoplasmic divisions to produce small polar bodies and ova. The **secondary spermatocyte** (E) is a diploid cell which undergoes equal cytoplasmic divisions to form **spermatids** which subsequently differentiate into **spermatozoa** during **spermiogenesis**.

    Gametogenesis is the process of formation of the gametes (ova and spermatozoa). This process occurs in the gonads (ovaries and testes). There are millions of gametes produced every day in males but usually only one every month in females. In males, meiosis is completed in the testes. In females, meiosis is not completed until the second polar body is cast off from the secondary oocyte, an even that is triggered by fertilization. The secondary oocyte is produced in the female just before ovulation, when the first polar body and a secondary oocyte are formed from a primary oocyte.

    Spermatogenesis is the process of production of spermatozoa. It occurs in the seminiferous tubules of the testes. Spermatogenesis begins at puberty and continues until death in the males, although there is a gradual decrease in sperm count with advanced age but spermatogenesis is a continuous process from puberty until death. Spermatogonia are present before puberty but they do not begin to differentiate into spermatozoa until puberty. Primary spermatocytes are tetraploid cells with paired bivalents (homologous chromosomes) in them. They form secondary spermatocytes (as a result of meiosis I) which in turn form haploid spermatids (as a result of meiosis II). Next, spermatids undergo a complex cytomorphogenesis (during spermiogenesis) to form spermatozoa.

33. When an oocyte is released from the ovary, all of the following structures are released into the lumen of the female reproductive tract **EXCEPT**:

(A) zona pellucida
(B) corona radiata
(C) theca lutein cells
(D) secondary oocyte
(E) first polar body

34. Which of the following **BEST** describes the acrosomal reaction?

(A) occurs in epididymis
(B) required for capacitation
(C) aids in degradation of extracellular matrix between corona radiata cells
(D) causes block to polyspermy
(E) prevents penetration of the zona pellucida

35. All of the following occur at fertilization **EXCEPT**:

(A) expulsion of first polar body
(B) establishment of diploid chromosome number
(C) block to polyspermy
(D) completion of second meiotic division in oocyte
(E) initiation of development

36. Fertilization is thought to occur normally in which location in the female?

(A) uterus
(B) cervix
(C) intramural part of the uterine tube
(D) ampullary part of the uterine tube
(E) vagina

37. Fertilization normally occurs within what time period after ovulation?

(A) 1 hour
(B) 6 hours
(C) 1 day
(D) 5 days
(E) 2 weeks

ANSWERS AND TUTORIAL ON ITEMS 33-37

The answers are: **33-C; 34-C; 35-A; 36-D; 37-C**. During formation of the **mature antral follicle**, the first meiotic division is completed just before ovulation, resulting in the formation of the **secondary oocyte** and the **first polar body**. It is thought that fertilization normally occurs within about one day of ovulation. When spermatozoa reach the immediate environment of the primary oocyte (and associated corona radiata and zona pellucida), the **acrosome reaction** occurs, leading to the release of hydrolytic enzymes used to degrade the extracellular matrix between granulosa cells of the **corona radiata** and for lysis of the zona pellucida. When the tip of the acrosome contacts the primary oocyte, **cortical granules** are discharged that cause changes in the oocyte membrane rendering it unfertilizable by other spermatozoa (**block to polyspermy**). At **fertilization** the second meiotic division occurs, resulting in the formation of the **ovum** and the **second polar body**. A third polar body sometimes forms as a result of completion of meiosis II in the first polar body. Fertilization also initiates development and results in the **haploid male pronucleus** entering the cytoplasm of the ovum where it fuses with the **haploid female pronucleus**, to form the **diploid zygote nucleus**.

Items 38-43

38. Implantation normally occurs **MOST** frequently in which location?

   (A) posterior uterine fundus
   (B) distal uterine tube
   (C) the cervical os
   (D) lower uterine segment

39. Which tissue is **MOST** directly responsible for degradation of endometrial epithelium during implantation?

   (A) uterine stroma
   (B) uterine endometrial epithelium
   (C) epiblast
   (D) cytotrophoblast
   (E) syncytiotrophoblast

12

40. Human chorionic gonadotrophin (hCG) is secreted by which tissue?

(A) syncytiotrophoblast
(B) cytotrophoblast
(C) epiblast
(D) hypoblast
(E) endometrial epithelium

41. Which statement **BEST** describes the most direct role of hCG in the female reproductive cycle?

(A) It stimulates completion of meiosis.
(B) It stimulates endometrial proliferation.
(C) It maintains the progestational state.
(D) It causes regression of the corpus luteum.
(E) It stimulates endometrial secretion.

42. In humans, the implantation site:

(A) is partially engulfed in maternal tissue
(B) is partially surrounded by endometrial epithelium
(C) is stimulated to invade extensively by the decidual cell reaction
(D) is surrounded by epiblast
(E) has its invasiveness limited by the decidual cell reaction

43. Decidual cells are **BEST** described as

(A) partially fetal and partially maternal in origin
(B) completely derived from maternal endometrial epithelium
(C) derivatives of maternal blood vessels
(D) derivatives of syncytiotrophoblast
(E) derivatives of maternal endometrial stroma

ANSWERS AND TUTORIAL ON ITEMS 38-43

The answers are: **38-A; 39-E; 40-A; 41-C; 42-E; 43-E**. During **implantation**, the blastocyst, which was formerly floating freely in the lumen of the female reproductive tract, attaches to and invades the **endometrium**. Implantation occurs normally in the **posterior wall** of the uterine body.

The outer epithelial **trophoblast** of the blastocyst adheres specifically to endometrial epithelial cells and then differentiates into two layers, the **cytotrophoblast** and the

syncytiotrophoblast. The cytotrophoblast proliferates and daughter cells fuse to form the syncytiotrophoblast which degrades and invades first the endometrial epithelium and then the endometrial stroma. The syncytiotrophoblast also secretes an important glycoprotein hormone called human chorionic gonadotropin (**hCG**). This hormone stimulates the **corpus luteum** to secrete **progesterone** and thus maintains the progestational state. Without a corpus luteum during the first trimester, the pregnancy would fail.

In humans, implantation is **interstitial**, i.e., the developing conceptus becomes completely engulfed by maternal tissue. The syncytiotrophoblast invades maternal tissue by degrading it and eventually degrades maternal blood vessels as well. The surface of the syncytiotrophoblast increases dramatically, first by the formation of lacunae and later, in collaboration with the cytotrophoblast, fetal connective tissue, and fetal blood vessels, contributes to **placental villi**. The syncytiotrophoblast is the boundary between the fetal tissue and is bathed directly in maternal blood. Its invasiveness is limited by the formation of **decidual cells**. These differentiate from endometrial stroma and form a shell around the implantation site, limiting the invasiveness of the conceptus.

Items 44-47

44. Ectopic pregnancy is **MOST** commonly found in the

    (A)   abdominal cavity
    (B)   distal uterine tube
    (C)   lower uterine segment
    (D)   cervix
    (E)   intramural uterine tube

45. Placenta previa is **BEST** defined as

    (A)   placenta detached from wall of uterus
    (B)   ischemic and calcified placenta
    (C)   placenta with marginal insertion of umbilical cord
    (D)   placenta located partially or completely covering the internal cervical os
    (E)   placenta located in the lower uterine segment

46. All of the following are examples of ectopic pregnancy **EXCEPT**:

    (A)   tubal pregnancy
    (B)   abdominal pregnancy
    (C)   implantation site in the rectouterine pouch (of Douglas)
    (D)   ovarian implantation
    (E)   placenta previa

14

47.   All of the following are associated with ectopic pregnancy **EXCEPT**:

    (A)   missed menstrual period
    (B)   pelvic inflammatory disease
    (C)   previous pelvic surgery
    (D)   absence of hCG in urine and blood
    (E)   shock

ANSWERS AND TUTORIAL ON ITEMS 44-47

The answers are: **44-B; 45-D; 46-E; 47-D**. **Ectopic pregnancy** is defined as an implantation site outside the lumen of the uterine cavity. More than 90% of all cases of ectopic pregnancy are found in the **distal part of the uterine tubes**. Inefficient tubal transport of the blastocyst is probably the cause of ectopic pregnancy. Previous history of pelvic inflammatory disease or pelvic surgery will increase the likelihood of adhesions and inflammation occurring in the tubes and thus decrease their transport capability. Since the early conceptus secretes hCG in any location, all of the typical signs of pregnancy, i.e. missed menstrual period and elevated hCG titer will occur. If the tubal pregnancy were to rupture, extensive bleeding and shock would be likely. **Figure 1.4** is a surgical specimen removed from a 19-year-old woman seen in the emergency room for right lower quadrant pain. She was found to have reduced hematocrit, her last menstrual period was 50 days earlier, and she had a positive pregnancy test. She had an ectopic pregnancy in her uterine tube.

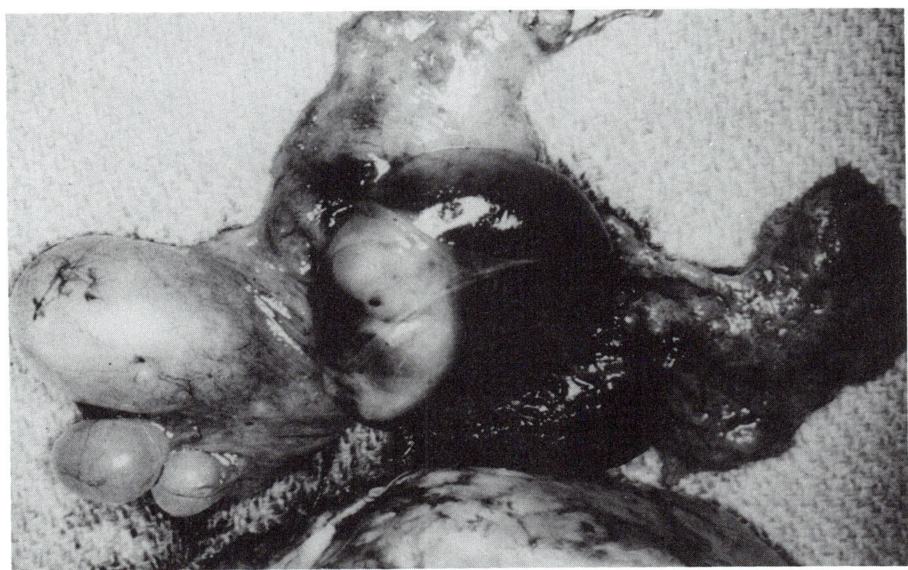

Figure 1.4

Although **placenta previa** (placenta formed partially or completely covering the internal cervical os) is an abnormal implantation site, it is not an ectopic pregnancy because the placenta and fetus are found completely within the uterus.

<u>**Items 48-52**</u>

48.   Which of the following is an ectodermal derivative?

      (A)   dermal fibroblast
      (B)   sweat gland
      (C)   skeletal muscle
      (D)   erythrocyte
      (E)   endothelium

49.   Which of the following is a mesodermal derivative?

      (A)   dermis
      (B)   epidermis
      (C)   enamel
      (D)   neuron in brain
      (E)   retina

50.   Which of the following is an endodermal derivative?

      (A)   endothelium
      (B)   endometrium
      (C)   tracheal epithelium
      (D)   melanocytes
      (E)   macrophages

51.   Which of the following is a neural crest derivative?

      (A)   oligodendrocyte
      (B)   adrenal cortex
      (C)   lymphocyte
      (D)   Schwann cell
      (E)   glucagon-secreting cell in islets of Langerhans

52. Which structure represents the cranial boundary between ectodermally derived epithelium and endodermally derived epithelium?

(A) lip
(B) stomodeum
(C) thyroglossal duct
(D) septum transversum
(E) respiratory diverticulum

ANSWERS AND TUTORIAL ON ITEMS 48-52

The answers are: **48-B; 49-A; 50-C; 51-D; 52-B**. During early development the **embryoblast** (inner cell mass) becomes divided into two layers, the **epiblast** and the **hypoblast**. During gastrulation three primary germ layers, **ectoderm**, **mesoderm**, and **endoderm** are formed. During neurulation the ectoderm forms a neural plate in the dorsal midline which folds up into a neural tube during neurulation. Just prior to the fusion of the neural folds dorsally, certain neuroectodermal cells leave the neuroepithelium and move to scattered and sometimes distant sites as the embryo enlarges. This population of cells is known as the **neural crest**. The remaining ectoderm, i.e. the ectoderm that forms neither the neural tube nor the neural crest, is called surface ectoderm.

The **ectoderm** forms the neuroectoderm and the surface ectoderm. The **neuroectoderm** forms the **neural tube** and **neural crest**. The neural tube forms the brain and spinal cord of the central nervous system (CNS). **Surface ectoderm** forms the outer surface epithelium of skin (epidermis) and its appendages, i.e., hair, hair follicles, nails, nail beds, sweat and sebaceous glands, and the exocrine portion of the mammary glands. Surface ectoderm also forms the corneal epithelium, the epithelial lining of the oral cavity in front the stomodeum (and thus the enamel of teeth and parts of some salivary glands), and the lining of the anal canal below the proctodeum.

Neural crest forms the neurons and glial cells in spinal ganglia, the myenteric plexus, and parasympathetic and sympathetic ganglia. In addition, it forms the dentin secreting odontoblasts of teeth, melanocytes, Schwann cells (PNS myelinating cells), the adrenal medulla, and pia-arachnoid meninges. The neural crest makes an important contribution to head mesenchyme and thus contributes to pharyngeal arch cartilage, bone, connective tissue, and the pupillary muscles and ciliary muscles in the eye. It also forms calcitonin-secreting parafollicular cells in the thyroid gland, parts of the cardiac outflow tracts (e.g., the truncoconal septum), and some enteroendocrine cells.

Mesoderm forms the entire cardiovascular system and blood. Most connective tissues (general and special) are also mesodermal derivatives. Therefore, except for those formed by neural crest, all other bones, cartilages, ligaments, tendons, fascia, and aponeuroses are formed from mesoderm. The vast bulk of skeletal and smooth muscle in the body is formed by mesoderm (except for the pupillary muscles, which are derived from the neural crest). Most of the urogenital system also arises from mesoderm.

Endoderm forms the epithelial lining of the gastrointestinal tract and its diverticula including the salivary glands, lungs, thyroid gland (follicular epithelial cells only), liver, pancreas, and gall bladder and their ducts. The transitional epithelium lining the urinary bladder is also endodermally derived.

Items 53-60

Match the embryonic structure from the list of answers below with the **MOST** accurate description of its developmental fate in the items below. Answers may be used once, more than once, or not at all.

(A)   Head fold of the embryo
(B)   Tail fold of the embryo
(C)   Somatopleuric intraembryonic mesoderm
(D)   Splanchnopleuric intraembryonic mesoderm
(E)   Intermediate mesoderm
(F)   Somite
(G)   Stomodeum
(H)   Proctodeum
(I)   Yolk sac
(J)   Yolk stalk (vitelline duct)
(K)   Septum transversum

53. Meckel's diverticulum is caused by persistence of this structure.

54. Forms enteric mural smooth muscle.

55. Growth causes heart primordium to move from a position cranial to the brain anlage.

56. Forms the ovary and metanephros.

57. Source of primordial germ cells.

58. Forms the earliest site of hematopoiesis.

59. Makes a major contribution to diaphragm and connective tissue of liver.

18

60.   All of the following are true for Meckel's diverticulum **EXCEPT**:

   (A)   more common in females than in males
   (B)   may contain ectopic gastric tissue
   (C)   most often found in the ileum
   (D)   can be attached to umbilicus by fibrous band
   (E)   found in about 1% of cadavers

ANSWERS AND TUTORIAL ON ITEMS 53-60

The answers are: **53-J; 54-D; 55-A; 56-E; 57-I; 58-I; 59-K; 60-A**. The growth of the neural
tube is an important feature in the establishment of the basic vertebrate body plan in humans.
The **head fold of the embryo** (A) shapes the head and moves the **stomodeum** (G) and
cardiac primordium from a dorsal cranial position to a ventral (rotated) position caudal to
the forebrain. Similarly, the **tail fold of the embryo** (B) moves the **proctodeum** (H)
(precursor of the lower anal canal) under the caudal end of the embryo. The **lateral folds of
the embryo** move the lateral plate mesoderm ventrally, thereby forming the tubular gut. The
body folds convert the yolk sac ventral to the gut into the **yolk stalk** (J) that courses
ventrally into the **yolk sac** (I). The yolk sac initially remains connected to the midgut by the
yolk stalk until the tenth week when it usually disappears. Persistent yolk stalk remnants can
lead to formation of **Meckel's diverticulum**. This anomaly is seen in about 1% of adults,
most commonly as a small diverticulum in the ileum and more commonly in males than in
females. It often contains ectopic gastric tissue and can lead to ileal ulceration. During its
early development, the yolk sac serves both as a source for primordial germ cells and as the
earliest site of hematopoiesis.

Intraembryonic mesoderm is divided into **paraxial mesoderm** that becomes
segmentally arranged into **somites** (F). It continues laterally as a segmentally arranged mass
of **intermediate mesoderm** (E). The intermediate mesoderm then leads into **lateral plate
mesoderm**. Somites become divided into **sclerotomes** (precursors of vertebral bodies and the
annulus fibrosus) and **dermomyotomes** (precursors of segmental dermis and skeletal muscle).
Intermediate mesoderm forms most of the urogenital system. The intraembryonic coelom
forms by coalescence of small clefts in the lateral plate which splits it into an outer
(superficial) **somatopleuric (somatic) mesoderm** (C) and an inner (deep) **splanchnopleuric
(splanchnic) mesoderm** (D). The somatic mesoderm forms body wall structures, while the
splanchnic mesoderm forms smooth muscle and connective tissue surrounding visceral
organs, e.g., enteric smooth muscle and cardiac muscle.

The **septum transversum** (K) is a mass of mesoderm that lies caudal to the heart
primordium but cranial to the vitelline duct. It makes a major contribution to the formation
of the diaphragm. The liver diverticulum grows into the septum transversum, eventually
forming liver parenchymal cells. The mesenchymal tissue in the septum transversum forms
blood vessels and connective tissue in the liver.

Items 61-70

For each labeled portion of a bone developing by endochondral ossification shown in **Figure 2.1**, choose the **MOST** appropriate descriptive phrase in the items below. Answers may be used once, more than once, or not at all.

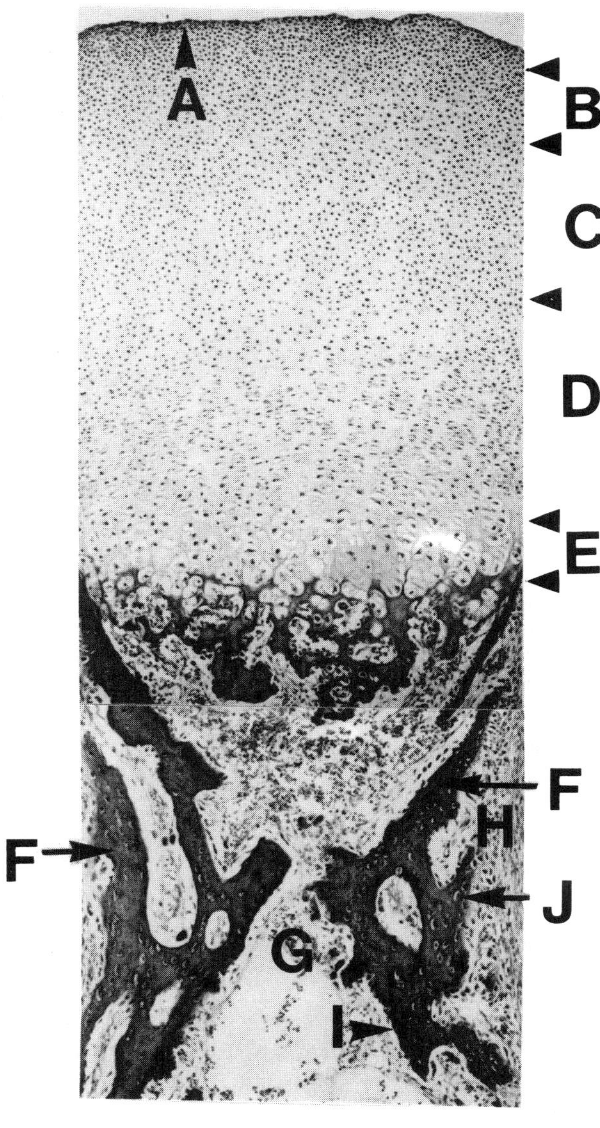

Figure 2.1

61.   This structure is a bony collar of calcified tissue.

62.   This structure is a zone of active mitosis in differentiated chondrocytes.

63.   This structure is a zone where the cartilaginous matrix is calcifying and chondrocytes
        are degenerating.

64.   This structure is a zone of where chondrocytes are undergoing hypertrophy.

65.   This structure will eventually become the periosteum of the long bone.

66.   This region ossifies first during development of a long bone.

67.   This structure contains mesenchymal cells that will eventually become hematopoietic
        stem cells.

68.   In the adult, this portion of the cartilaginous model persists as smooth hyaline
        cartilage.

69.   Mesenchymal cells here will form the endosteum.

70.   This structure is an osteocyte lacuna.

ANSWERS AND TUTORIAL ON ITEMS 61-70

The answers are: **61-F; 62-C; 63-E; 64-D; 65-H; 66-F; 67-G; 68-A; 69-I; 70-J**. **Figure 2.1**
is a photomicrograph of the development (by classical **endochondral bone formation**) of one
of the phalanges in a feline embryo. This bone, like most other bones in the appendicular
skeleton, forms by ossification of a preformed cartilaginous model. The future **articular
surface** (A) is the only remaining remnant of the cartilage model persisting in the adult. The
cartilage grows interstitially and appositionally. Chondrocytes is the **resting zone** (B) can
become mitotically active. Soon after formation, perhaps due to limitation of nutrient
diffusion by the **bony collar** (F), the **zone of proliferation** (C) gives way to a **zone of
hypertrophy** (D) and then a **zone of degeneration and calcification** (E). Limb bud
mesenchyme peripheral to the developing bone model forms a compact layer of **periosteal
fibroblasts** (H) and hematopoietic stem cells in the future **marrow cavity** (G). Other
mesenchymal cells on the inner surfaces of trabeculae of bone form the **endosteum** (I), an
osteogenic epithelioid layer of osteoblasts. **Lacunae** (J) contain entrapped osteocytes which
communicate by osteocyte processes within minute canaliculi (not shown).

<u>**Items 71-72**</u>

Choose the **BEST** response.

71. All of the following bones are formed at least in part by intramembranous ossification
     EXCEPT:

     (A) frontal bone
     (B) maxilla
     (C) mandible
     (D) malleus
     (E) temporal bone

72. All of the following bones are formed at least in part by endochondral ossification
     EXCEPT:

     (A) temporal bone
     (B) occipital bone
     (C) zygomatic arch
     (D) tibia
     (E) radius

ANSWERS AND TUTORIAL ON ITEMS 71 AND 72

The answers are: **71-D; 72-C. Intramembranous ossification** involves direct deposition of
osteoid by osteoblasts followed by calcification of the osteoid. This process proceeds
spontaneously in mesenchyme without the formation of a cartilaginous model first. The bones
of the vault of the skull (e.g., frontal, parietal, and part of occipital and temporal bones) and
many facial bones (e.g., zygomatic arch, nasal, maxilla, and mandible) form this way. In
addition, most of the clavicle forms by intramembranous ossification.

    Endochondral ossification requires the production and then destructive ossification of
a cartilaginous model of the bone. Otherwise, it proceeds much like intramembranous
ossification. Most bones in the body form by endochondral ossification including those of the
limbs, vertebral bodies, ribs, pelvis and shoulder girdle, and base of the skull.

Items 73-79

Match the clinical entity or situation caused by a structural defect in the items below with the portion of the embryonic mesoderm **MOST** responsible for this abnormality. Answers may be used once, more than once, or not at all.

    (A)   Somatopleuric mesoderm
    (B)   Splanchnopleuric mesoderm
    (C)   Sclerotome
    (D)   Dermomyotome
    (E)   Intermediate mesoderm
    (F)   Notochord
    (G)   None of the above

73. Prune belly

74. Congenital diaphragmatic hernia

75. Meckel's diverticulum

76. Renal agenesis

77. Nucleus pulposus pressing on sciatic nerve root secondary to traumatic tear in annulus fibrosus

78. Congenital absence of vertebral body

79. Horseshoe kidney

ANSWERS AND TUTORIAL ON ITEMS 73-79

The answers are: **73-D; 74-A; 75-G; 76-E; 77-F; 78-C; 79-E**. **Prune belly** is due to a congenital absence of the abdominal wall musculature which is derived from the regional dermomyotomes. It is 20 times more common in males than females and is often associated with urogenital anomalies. It may be caused by abnormal distention of the abdomen which secondarily prevents development of abdominal musculature.

Congenital diaphragmatic hernia is due to failure of ventral growth and fusion of the pleuroperitoneal membrane (most often on the left side) with the septum transversum. The pleuroperitoneal membrane is a somatopleuric mesoderm derivative.

Meckel's diverticulum is an ileal diverticulum which is a remnant of the vitelline duct (i.e., none of the above). Normally, this embryonic connection between the midgut and the yolk sac is obliterated.

Renal agenesis is due to failed reciprocal inductive interactions between the ureteric bud and the metanephric blastema, both intermediate mesoderm derivatives. Horseshoe kidney occurs when the caudal poles of the metanephroi fuse. Cranial ascent of the fused metanephroi is blocked by the inferior mesenteric artery.

The **nucleus pulposus**, that gelatinous cushion surrounded by the annulus fibrosus in the intervertebral discs, is derived from the notochord. Indeed, it is the only known persistent notochordal derivative in the adult. The notochord in the center of developing vertebral bodies disappears with ossification. When the intervertebral disc is traumatically torn, often at the L4,5 level, weight on the lumbar vertebral bodies expresses the semi-liquid nucleus pulposus onto the sciatic nerve roots, causing pain and possibly loss of sensory and motor function.

Vertebral bodies can be entirely missing or may be misshapen. They are intersegmentally derived from **sclerotomes**.

<u>Items 80-82</u>

Choose the **BEST** response.

80.	Which of the following is a dermomyotome derivative?

    (A)	spinal ganglia
    (B)	vertebral bodies
    (C)	epidermis
    (D)	intervertebral discs
    (E)	dense irregular connective tissue of skin

81.	Which of the following is a sclerotome derivative?

    (A)	vertebral body
    (B)	nucleus pulposus
    (C)	dermal fibroblasts
    (D)	spinal ganglia
    (E)	limb bud musculature

82.   Which of the following is a splanchnopleuric mesoderm derivative?

    (A)   pleuroperitoneal membrane
    (B)   intercostal muscles
    (C)   smooth muscle in the wall of the stomach
    (D)   teeth
    (E)   bones in the extremities

ANSWERS AND TUTORIAL ON ITEMS 80-82

The answers are: **80-E; 81-A; 82-C**. Soon after its formation, the intraembryonic mesoderm becomes subdivided into paired paraxial **somites**, **intermediate mesoderm**, and **lateral plate mesoderm**. Each somite later forms a **sclerotome** (vertebral bodies, annulus fibrosus of intervertebral discs) and a **dermomyotome**. The dermomyotome then forms dermal dense irregular connective tissues of the dermis and segmentally arranged myotomes. The myotomes later form skeletal muscles. Intermediate mesoderm forms most of the urogenital system. Lateral plate mesoderm becomes divided (by the intraembryonic coelom) into **somatopleuric mesoderm** and **splanchnopleuric mesoderm**. The former forms body wall structures. The latter forms mesodermal derivatives closely associated with the gastrointestinal tract and its diverticula. For example, the cartilage in the trachea and the smooth muscle in the wall of the stomach are splanchnic mesodermal derivatives.

Items 83-88

A child born to apparently normal parents is unusually short with a depressed nasal bridge, a prominent forehead, and significant bowing of the lower limbs. Further tests reveal that the child is affected by achondroplasia.

83.   In **MOST** cases, this developmental defect is caused by which fundamental underlying genetic mechanism?

    (A)   aneuploidy
    (B)   autosomal dominant mutation
    (C)   autosomal recessive mutation
    (D)   trisomy
    (E)   sex-linked mutation

84.   If a mother and father with this condition had a child, the probability of this child being unaffected by this condition would be

(A)   0%
(B)   25%
(C)   50%
(D)   75%
(E)   100%

85.   The incidence of this condition per live birth is approximately

(A)   1 in 1,000
(B)   1 in 2,000
(C)   1 in 5,000
(D)   1 in 10,000
(E)   1 in 100,000

86.   Which developmental process is **MOST** disrupted in this anomaly?

(A)   somitogenesis
(B)   osteoid secretion
(C)   neural crest cell migration
(D)   chondrocyte proliferation at epiphyseal plate
(E)   intramembranous ossification

87.   All of the following characteristics are common in female achondroplastic patients **EXCEPT**:

(A)   severely reduced height due to diminished growth of long bones
(B)   normal levels of growth hormone
(C)   infertility
(D)   normal mental capacity
(E)   difficult labor and delivery

88.   The cause of achondroplasia is

(A)   hyposecretion of growth hormone
(B)   lack of growth hormone receptors
(C)   mutation in gene encoding somatostatin
(D)   mutation in gene encoding fibroblast growth factor-3 receptor
(E)   unknown

The answers are: **83-B; 84-B; 85-D; 86-D; 87-C; 88-D**. **Achondroplasia** is the most common chondrodystrophy (incidence approximately 1 in 10,000 live births). It is caused by an **autosomal dominant mutation**. Most cases arise as spontaneous new mutations with parents unaffected by achondroplasia. The heterozygous form is compatible with a long life and normal intelligence. The homozygous form is usually lethal early in life. If heterozygous parents were to have a child, it would have a 25% chance of being unaffected by achondroplasia. The remaining offspring would be 50% heterozygous affected and 25% homozygous (lethal).

Recent studies have shown that most cases of achondroplasia are due to a mutation in the gene encoding **fibroblast growth factor-3 receptor**. Absence of this receptor results in epiphyseal plates that do not show normal proliferative activity and thus many of the bones formed by endochondral ossification (e.g., bones in the extremities and base of skull) are abnormally short. The nasal bridge depression is due to abnormal development of the base of the skull. Adults with achondroplasia have normal mentation and are fertile. Pregnant achondroplastics are routinely delivered by cesarean section due to **cephalopelvic disproportion**.

Items 89-90

Choose the **BEST** response

89.   All of the following are true statements concerning limb buds **EXCEPT**:

    (A)   apical ectodermal ridge induces mesenchymal differentiation
    (B)   limb bud mesenchyme forms bone and ligaments
    (C)   cell death accompanies digit formation
    (D)   the forelimb buds arise after the hindlimb buds
    (E)   syndactyly is caused by abnormal limb bud development

90.   Which statement **BEST** characterizes the role of limb bud mesenchyme in limb development?

    (A)   source of skeletal muscles but not bones
    (B)   source of bones but not skeletal muscles
    (C)   little cell death involved in limb morphogenesis
    (D)   mesenchyme programmed to form proximal structures before distal structures
    (E)   limb develops in a distal-proximal sequence and posterior-anterior sequence

ANSWERS AND TUTORIAL ON ITEMS 89-90

The answers are: **89-D; 90-D**. The **apical ectodermal ridge** induces generic differentiation of the **limb bud mesenchyme**, which gives rise to the tendons and bones of the limb. There is a timing mechanism operative during limb development, such that proximal structures are specified before distal structures. Limb skeletal musculature also arises locally from limb bud mesenchyme. There is a cranial to caudal sequence of limb bud development. The sculpting of the digits is caused by **apoptosis** of tissue between them and when it fails, syndactyl can result. **Hox gene** expression in limb morphogenesis occurs in an anterior to posterior and a proximal to distal sequence.

Items 91-94

A newborn male infant is delivered prematurely and alive. The infant had a beak-like nose, small thorax, and numerous fractures in the limbs. One day after birth, the child died from respiratory failure.

91.   The **MOST** likely diagnosis is

    (A)   osteogenesis imperfecta, type I
    (B)   osteogenesis imperfecta, type II
    (C)   achondroplasia
    (D)   Marfan's syndrome
    (E)   Hurler's syndrome

92.   The incidence of this condition is approximately

    (A)   1 in 5,000
    (B)   1 in 10,000
    (C)   1 in 50,000
    (D)   1 in 100,000
    (E)   1 in 500,000

93.   The mode of genetic transmission of this condition is

    (A)   sex-linked recessive
    (B)   sex-linked dominant
    (C)   autosomal dominant
    (D)   autosomal recessive
    (E)   aneuploidy

94. The biochemical lesion in this disease affects which important constituent of the extracellular matrix of connective tissue in limbs?

(A)   hyaluronic acid
(B)   type I collagen
(C)   type IV collagen
(D)   chondroitin sulfate
(E)   heparan sulfate

ANSWERS AND TUTORIAL ON ITEMS 91-94

The answers are: **91-B; 92-C; 93-C; 94-B**. **Osteogenesis imperfecta** (OI) is a complex group of diseases all associated with increased fragility of bones due to mutations in the genes for the polypeptides comprising **Type I collagen**, the predominant collagen of bone. This child was afflicted by a severe variety, **OI Type II**. This form, unlike the less severe OI Type I, is almost always fatal soon after birth. The most acute problems associated with OI Type II are breathing difficulties secondary to severe reduction in the thorax and possible damage to the brain from reduced ossification of the vault of the skull.

This rare disease, whose incidence is variously stated as about 1 in 20,000 to 1 in 60,000 infants (depending on the authority cited), is caused by **autosomal dominant mutations** (most often spontaneous new mutations) that alter the structure of Type I collagen. There is often severe reduction in secretion of Type I collagen. The tensile strength of bone is partially due to the presence of numerous Type I collagen fibrils. Thus, with severe decreases in Type I collagen in the extracellular matrix of bone, there is invariably a concomitant increase in bone fragility.

Items 95-97

Choose the **BEST** response

95. In a normal, 10-year-old, female child, the bulk of growth in length of long bones is caused by

(A)   mitosis of periosteal osteoblasts
(B)   interstitial growth of bones
(C)   proliferation of chondrocytes in the epiphyseal plate followed by ossification
(D)   proliferation of chondrocytes in the articular cartilage
(E)   proliferation of endosteal osteoblasts

30

96. In a normal, 25-year-old adult, growth in length of long bones is normally precluded because

(A) epiphyseal plates have reduced sensitivity to growth hormone
(B) epiphyseal plates stop growing and become completely ossified
(C) osteolysis outstrips osteogenesis
(D) osteoclastic activity becomes more intense
(E) osteoblasts disappear

97. In the repair of a simple fracture of the radius, which cell type is **MOST** active in the secretion of osteoid in the repairing callus?

(A) osteoclast
(B) fibroblast
(C) monocyte
(D) osteoblast
(E) osteoprogenitor cell

ANSWERS AND TUTORIAL ON ITEMS 95-97

The answers are: **95-C; 96-B; 97-D**. Bones are dynamic organs with extensive developmental potential during childhood. Even after puberty, when growth in the length of bones is no longer possible because of ossification of the **epiphyseal plate**, bones continue to have a latent capacity for reformation during fracture repair. In addition, the calcified matrix of bony tissue is constantly turning over as part of the calcium regulatory system in the body.

The growth in length of long bones during childhood is primarily due to proliferation, hypertrophy, and calcification of cartilage in the epiphyseal plate. This mechanism is accelerated during the pubertal growth spurt and then ceases after puberty due to the complete ossification of the epiphyseal plate.

In the normal steady state, bone resorption by osteocytes and osteoclasts (to release calcium into the blood) is balanced by bone deposition by osteoblasts (to sequester calcium from the blood). There is constant balanced deposition and resorption of bone. When a bone is fractured, osteoblasts are stimulated to secrete osteoid at the fracture site. This osteoid then calcifies to form a new **bony callus** which repairs the fracture. With advancing age, especially in post-menopausal women, bone deposition is outstripped by bone resorption, leading to **osteoporosis**.

For each description of a bone of the skull in the items below, choose the **MOST** appropriate labeled area in **Figure 2.2**. Answers may be used once, more than once, or not at all.

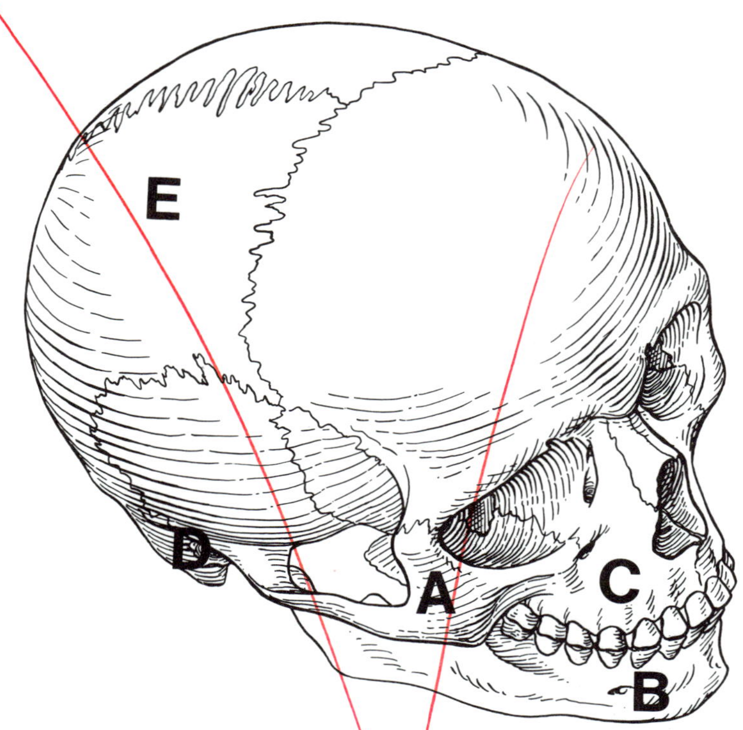

Figure 2.2

98. This bone is formed by intramembranous ossification and is part of the membranous viscerocranium. It contains no odontoblast derivatives.

99. This bone is formed by endochondral ossification in its inferior part and intramembranous in its superior part.

100. This bone is part of the membranous viscerocranium and is derived from the maxillary component of the first pharyngeal arch. The palatine processes form on its medial border.

101. This bone is part of the membranous neurocranium and the anterior fontanelle forms on its anterior border.

102. This bone forms by intramembranous ossification around Meckel's cartilage.

ANSWERS AND TUTORIAL ON ITEMS 98-102

The answers are: **98-A; 99-D; 100-C; 101-E; 102-B**. The **maxilla** (C) and **mandible** (B) are both parts of the **membranous viscerocranium**, as is the **zygomatic arch** (A). The mandible forms around Meckel's cartilage. The maxilla has palatine processes on its medial borders and these form most of the hard palate. Both the mandible and the maxilla have teeth with dentin (an odontoblast derivative) but the zygomatic arch does not. The **parietal bone** (E) is part of the **membranous neurocranium** and it forms the posterior border of the anterior fontanelle. The deep (petrous) portion of the **temporal bone** (D) is formed in a cartilaginous model and the superficial (squamous) portion is formed by intramembranous ossification.

Items 103-110

Experimental studies in laboratory animals have revealed a good deal about the molecular mechanisms of limb development. Match the developmental entity in the answers below with the **MOST** appropriate descriptive phrase in the items below. Answers may be used once, more than once, or not at all.

(A)   Apical ectodermal ridge
(B)   Nonapical ectodermal ridge ectoderm
(C)   Cervical and lumbar somites
(D)   Limb bud mesenchyme
(E)   Apoptosis
(F)   Hox-4
(G)   Retinoic acid
(H)   Zone of polarizing activity

103.   Responsible for formation of spaces between digits.

104.   A gradient of this regulates craniocaudal structure of limb.

105.   Source of putative morphogen that regulates craniocaudal limb structure.

106.   Produce transcription factors that are expressed in a craniocaudal sequence.

107.   Induce somatopleuric mesoderm to form limb buds.

108.   Responsible for induction of apical ectodermal ridge.

109.   Responsible for instructing nonspecific differentiation of limb bud mesenchyme.

110.   Age of this structure determines formation of proximal or distal limb bones.

ANSWERS AND TUTORIAL ON ITEMS 103-110

The answers are: **103-E; 104-G; 105-H; 106-F; 107-C; 108-D; 109-A; 110-D**. The limb buds develop in a cranial to caudal sequence. The flank region has the general ability to form limb buds. The specific location of limb buds in the cervical and lumbar regions is due to the inductive stimulation of the somatopleure by cervical (forelimb) and lumbar (hindlimb) **somites**.

Once formed by somitic induction, the limb buds consist of an **apical ectodermal ridge** (AER), nonapical ectodermal ridge cells, and a deep mass of mesenchymally derived fibroblasts. The AER is induced to form by the underlying limb bud mesenchyme. The AER in turn is essential for inducing further mesenchymal differentiation, although it functions in a nonspecific or generic fashion. The other ectodermal cells form the epidermis and associated appendages (nails, hair, sweat and sebaceous glands) on the limb. The **limb bud mesenchyme** forms cartilaginous models of bones and then bones, as well as tendons, ligaments and other connective tissue elements in the limb. Limb musculature arise within the limb bud from the limb bud mesenchyme.

On the proximal, caudal margin of the limb bud, there is a **zone of polarizing activity** (ZPA). Although this is a controversial topic, some research suggests that the ZPA elaborates **retinoic acid**, a potent morphogen that defines the normal craniocaudal sequence of the limb. Perhaps this explains why most of us have the same digit sequences on our hands and feet. In mice, homeobox genes of the **Hox-4** family (Hox-4.4 through Hox-4.7) are expressed sequentially in a craniocaudal (and proximodistal) sequence. There is evidence that retinoic acid can trigger precocious expression of certain Hox-4 genes. Digital rays are sculptured into digits by **apoptosis** or programmed cell death.

34

CHAPTER III
DEVELOPMENT OF THE NERVOUS
AND ENDOCRINE SYSTEMS

<u>Items 111-115</u>

Choose the **BEST** response.

111.   All of the following are true for the proliferating neuroepithelium **EXCEPT**:

(A)   The internal limiting membrane is a basement membrane at the lumen.
(B)   The external limiting membrane is the basement membrane at the periphery of the neural tube.
(C)   It is a pseudostratified epithelium.
(D)   Interkinetic nuclear migration is responsible for the pseudostratified character of the epithelium.
(E)   Cytokinesis occurs near the internal limiting membrane.

112.   Which of the following is **TRUE** for the glia limitans?

(A)   It contains no glial cells.
(B)   Few cells here have foot processes on the external limiting membrane.
(C)   It forms the outer boundary of neural tube derivatives.
(D)   It has numerous microglia.
(E)   It is part of the dura mater.

113.   Which of the following is a neuroblast derivative?

(A)   oligodendrocyte in the spinal cord
(B)   astrocyte in the spinal cord
(C)   Schwann cell in a peripheral nerve fiber
(D)   rod in the retina
(E)   pigmented epithelial cell in the retina

114.   All of the following statements are true for glioblasts **EXCEPT**:

    (A)   They form astrocytes.
    (B)   The form oligodendroglial cells.
    (C)   They form myelinating cells in the central nervous system.
    (D)   Many arise as a result of mitotic activity in the neuroepithelium.
    (E)   They form Schwann cells in a peripheral nerve fiber.

115.   Which of the following statements is **TRUE** for myelination?

    (A)   Neural crest derivatives accomplish it in the CNS.
    (B)   Oligodendroglial cells accomplish it in the PNS.
    (C)   The mesaxon grows to form multiple layers of membrane.
    (D)   It is completed by birth.
    (E)   It is most extensive in the CNS grey matter.

ANSWERS AND TUTORIALS ON ITEMS 111-115

The answers are: **111-A; 112-C; 113-D; 114-E; 115-C**. The **internal limiting membrane** is located at the lumen of the neural tube but it is not really a membrane. Instead, it consists of apical tight junctions between neuroepithelial cells. The **external limiting membrane** is located at the periphery of the neural tube and is the basement membrane for the neuroepithelium. The neuroepithelium is a pseudostratified epithelium and this characteristic is due to **interkinetic nuclear migration**.

The **glia limitans** forms at the external limiting membrane as astrocyte foot processes become pushed up against the external limiting membrane from within the neuroepithelium. Astrocytes (a kind of glial cell) are definitely part of the glia limitans. It forms the outer boundary of the neural tube derivatives.

Neuroblasts form many different types of neurons in the nervous system, including the cones and rods in the neural retina. Glioblasts form glial cells such as oligodendroglial cells and astrocytes. Schwann cells are neural crest derivatives responsible for PNS myelination. The neuroblasts are capable of extensive mitotic divisions after they are born. For example, in the cerebral and cerebellar cortex, an entire population of neuroblasts divides extensively to form the peripheral grey matter of the brain.

Glioblasts form several different kinds of glial cells including astrocytes and oligodendroglial cells. The latter cells are responsible for CNS myelination. The microglia are part of the mononuclear phagocyte system and arise from monocytes rather than the neuroepithelium, which is the source of glioblasts and all other glial cells.

36

Myelination is the result of the activity of **Schwann cells** (neural crest derivatives) in the PNS and **oligodendroglial cells** (glioblast derivatives) in the CNS . The mesaxon of the myelinating cells grows extensively in a spiral fashion around processes (axons and dendrites) to be myelinated. Myelination occurs before and after birth and is more prominent in white matter than in grey matter.

Items 116 and 117

Choose the **BEST** response

116.   Which of the following are alar plate (lamina) derivatives?

    (A)   spinal cord sensory neurons
    (B)   spinal cord motor neurons
    (C)   general visceral efferent nuclei
    (D)   spinal ganglia

117.   Which of the following are basal plate (lamina) derivatives?

    (A)   spinal cord sensory neurons
    (B)   spinal cord motor neurons
    (C)   general visceral afferent nuclei
    (D)   special visceral afferent nuclei
    (E)   spinal ganglia

ANSWERS AND TUTORIAL ON ITEMS 116 AND 117

The answers are: **116-A; 117-B**. The **alar plate (lamina)** is located dorsally in the neural tube. It contains the cell bodies of developing sensory neurons in the spinal cord and brain. The motor neurons develop in the basal plates. The neurons in spinal ganglia are neural crest derivatives. The sensory afferent brain nuclei arise from the alar plate whereas the motor efferent brain nuclei form in the basal plate.

The **basal plate (lamina)** is located ventrally in the neural tube. It contains the cell bodies for developing motor neurons in the spinal cord and brain. The tegmental nuclei in the mesencephalon and the general visceral efferent nuclei in the rhombencephalon form from the basal plate. The special visceral afferent nuclei in the rhombencephalon form from the alar plate.

Directions: The groups of questions below consist of lettered choices followed by several numbered items. For each numbered item select the **ONE** lettered choice with which it is **MOST** closely associated. Answers may be used once, more than once, or not at all.

Items 118-122

For each description of the development of an adult brain structure, select the **MOST** appropriate lettered adult brain structure in **Figure 3.1** below.

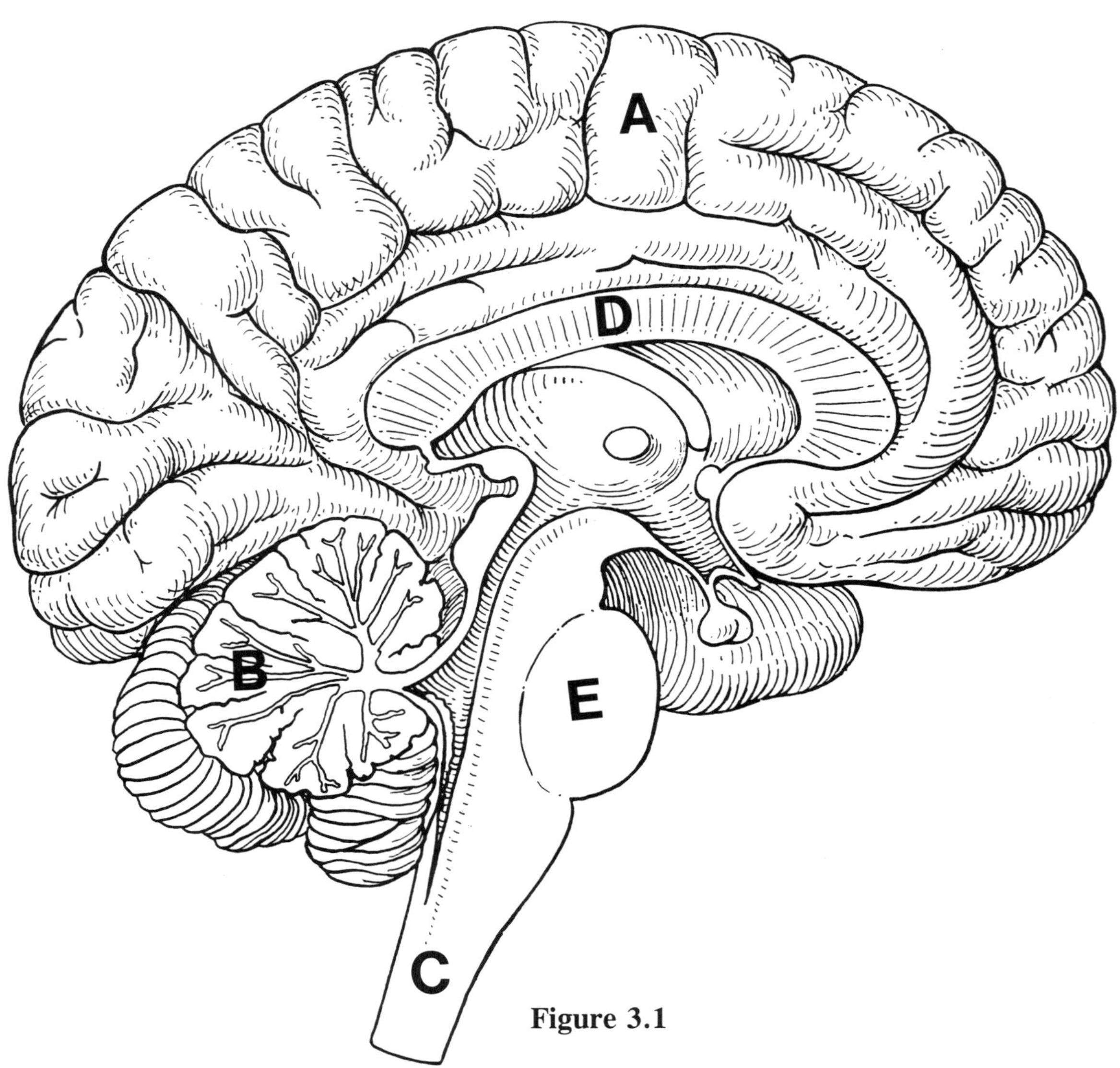

Figure 3.1

118.   This structure develops from the myelencephalic part of the rhombencephalon.

119.   This structure develops from the lateral alar plate expansions of the telecephalic part of the prosencephalon.

120.   This structure develops from the rhombic lip.

121.   This structure develops from the metencephalic part of the rhombencephalon but not from the rhombic lip.

122.   This structure contains commissural fibers between developing cerebral hemispheres.

Items 123-127

For each description of a structure in the developing spinal cord, select the **MOST** appropriate lettered structure shown in the photomicrograph in **Figure 3.2** below.

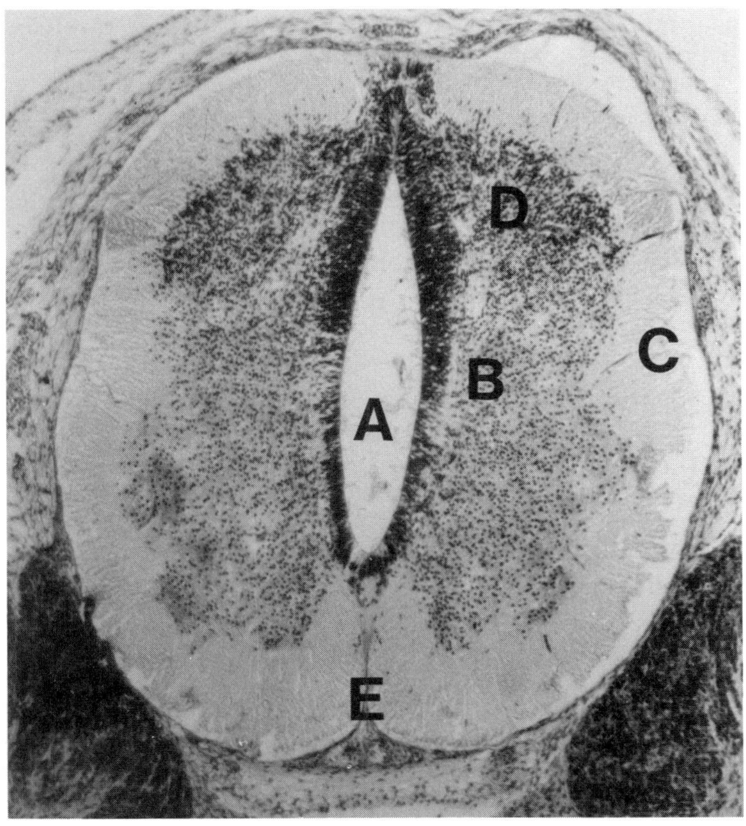

Figure 3.2

123. This structure is formed by rapid growth of motor neurons in the basal plate.

124. This structure was once continuous with the amniotic cavity but now communicates with the brain ventricles.

125. This structure contains neuroblasts that will differentiate into neurons which synapse with neural crest-derived neurons in the spinal ganglia.

126. This structure contains many neuronal cell bodies and relatively little myelin.

127. This structure contains relatively few neuronal cell bodies and many myelinated axons and dendrites.

ANSWERS AND TUTORIAL ON ITEMS 118-122

The answers are: **118-C; 119-A; 120-B; 121-E; 122-D**. The **cerebrum** (A) is formed by the telencephalic hemispheres which are in turn derived from the prosencephalon (forebrain). The **cerebellum** (B) forms in the rhombic lip of the metencephalon which is in turn derived from the rhombencephalon (hindbrain). The **medulla oblongata** (C) forms from the myelencephalon which in turn develops from the caudal portion of the rhombencephalon. The **corpus callosum** (D) is a major site of commissural fibers connecting the two halves of the developing cerebrum. The **pons** (E) forms from the metencephalic part of the rhombencephalon.

ANSWERS AND TUTORIAL ON ITEMS 123-127

The answers are: **123-E; 124-A; 125-D; 126-B; 127-C**. The **central canal** of the spinal cord (A) faced onto the **amniotic cavity** when the neural plate was flat. During neurulation, the neural plate rolls up into a tubular structure. Neuropores at the cranial and caudal extremes of the neural canal remain in open communication with the amniotic cavity for a time. Even after closure of the neuropores, the lumen of the neural tube is a continuous cavity from the ventricles of the brain to the central canal of the spinal cord. This cavity contains cerebrospinal fluid. The **mantle layer** (B) contains many neuroblasts but little myelin. The **marginal layer** (C) contains few neurons and many myelinated processes. The **alar plate** (D) (dorsally) contains neuroblasts that differentiate into sensory neurons which in turn synapse with the pseudounipolar neurons formed in the spinal ganglia (neural crest derivatives). The

ventral median fissure (E) is a deep recess that appears in the neural tube as a result of the relatively greater growth in the flanking basal plate. The **basal plate** (unlabeled) contains neuroblasts that differentiate into large motor neurons whose axons extend peripherally to innervate skeletal muscles.

Item 128

128.   Which of the following is the **MOST** accurate explanation of why the cauda equina forms?

   (A)   The spinal cord migrates cranially in the vertebral canal.
   (B)   The vertebral column elongates at a faster rate than the spinal cord.
   (C)   The lower spinal nerves grow faster than the spinal cord and push the cord cranially.
   (D)   The filum terminale is composed of elastic tissue thereby allowing the spinal cord to move cranially.
   (E)   The lower spinal ganglia do not anchor the lower part of the cord as well as those at higher levels do.

ANSWER AND TUTORIAL ON ITEM 128

The answer is: **128-B**. As the embryo/fetus enlarges the vertebral column naturally elongates. Because the spinal cord does not grow longitudinally as fast as the vertebral column, its caudal termination occurs at a higher and higher vertebral level. At the end of the embryonic period (8 weeks) the vertebral column and spinal cord are the same length with the spinal nerve roots coursing horizontally from the spinal cord. At the end of the second trimester (24 weeks), the spinal cord ends at the level of the S1 vertebra. At birth, it ends at the L3 vertebra and in the adult at the disc between the L1 and L2 vertebrae. As the distance between the termination of the spinal cord and vertebral column increases, the lower spinal nerve roots become progressively more vertical in their course to exit the vertebral column.

129. All of the following are neural crest derivatives **EXCEPT**:

    (A)   neurons in myenteric plexus
    (B)   neurons in spinal ganglia
    (C)   neurons sympathetic ganglia
    (D)   astrocytes
    (E)   Schwann cells

130. Which of the following are glioblast derivatives?

    (A)   ependymal cells
    (B)   Schwann cells
    (C)   microglia
    (D)   Purkinje cells in cerebellum
    (E)   fibrous astrocytes

131. All of the following are neuroblast derivatives **EXCEPT**:

    (A)   pyramidal neurons in the cerebrum
    (B)   oligodendroglial cells
    (C)   Purkinje cells in cerebellum
    (D)   sensory neurons in spinal cord
    (E)   motor neurons in spinal cord

132. **TRUE** statements concerning the central nervous system include which of the following?

    (A)   It is a highly modified epithelium.
    (B)   It contains mesodermal derivatives.
    (C)   It is surrounded by a basement membrane in the embryo.
    (D)   Cell division plays an important role in its morphogenesis.
    (E)   All of the above are true.

ANSWERS AND TUTORIAL ON ITEMS 129-132

The answers are: **129-D; 130-E; 131-B; 132-E**. The central nervous system is formed by the **neural tube** which in turn is derived from the **neural plate**. In all stages of its development, this neuroepithelium is associated with a basement membrane. During formation of the neural plate, surface ectoderm thickens into a pseudostratified epithelium and then rolls up into a tube. As the dorsal neural folds fuse in the midline, **neural crest** cells leave the neuroepithelium and become dispersed widely throughout the embryonic body. Extensive cell proliferation produces an increasingly thickening neural tube wall. Proliferative neuroepithelial cells differentiate into **neuroblasts** and **glioblasts**. The former differentiate into the many different types of neurons found in the central nervous system, e.g., Purkinje cells in the cerebellum, pyramidal neurons in the cerebrum, as well as motor and sensory neurons in the spinal cord. Many of these neuroblasts, e.g., in the ventral motor horns (basal plate derivatives) extend axons, sometimes at great distances, into the peripheral nervous system where they innervate end organs such as skeletal muscles. The glioblasts differentiate into the major glial cells such as the oligodendroglial cells and the astrocytes, and other less abundant glial cell types. The microglia are part of the mononuclear phagocyte system and are derived from the bone marrow (mesoderm).

The **neural crest** also makes an important contribution to the development of the nervous system. The Schwann cells are myelinating cells of the PNS and are neural crest derivatives. The neurons in the spinal (sensory) ganglia, and many sympathetic and parasympathetic ganglia (e.g., the myenteric plexus), and adrenal medullary cells are also neural crest derivatives.

For each anatomical or functional description of the role of a structure in the development of the nervous system in the questions, choose the **MOST** appropriate embryonic structure in the answers below. Answers may be used once, more than once, or not at all.

(A)   Marginal layer
(B)   Mantle layer
(C)   Ependymal layer
(D)   Lumen of neural tube
(E)   External limiting membrane
(F)   Internal limiting membrane
(G)   Roof plate
(H)   Floor plate
(I)   Alar plate
(J)   Basal plate

133.   This structure in the early neural tube contains many neuronal cell bodies. It is the precursor of all of the grey matter.

134.   This structure is the basement membrane for the neuroepithelium.

135.   This structure lines ventricles.

136.   This develops into the white matter which contains much myelin.

137.   This structure forms the central canal of the spinal cord and ventricles of the brain in the definitive central nervous system.

138.   This structure becomes highly modified to form the cuboidal epithelium of choroid plexus.

139.   This thin structure perforates in three locations, forming an outflow tract for cerebrospinal fluid.

140.   In laboratory animals, this structure has been shown to play an important role in induction of motor neurons.

141.   This structure contains neuroblasts that will eventually form a large collection of ventral horn motor neurons.

The answers are: **133-B; 134-E; 135-C; 136-A; 137-D; 138-C; 139-G; 140-H; 141-J**. The neural tube early in its development is a pseudostratified epithelium surrounded by a basement membrane which persists and is known as the **external limiting membrane** (E). At the lumen, apical tight junctions between neuroepithelial cells lead to the formation of the **internal limiting membrane** (F). The **lumen of the neural tube** (D) persists as the central canal of the spinal cord and the ventricles and aqueduct of the brain. During early development of the neural tube, epithelial cells round up and their nuclei migrate toward the lumen of the neural tube. After cell division, some of these cells remain as stem cells and other daughter cells become neuroblasts or glioblasts. Once the development of the neural tube is complete, there is an **ependymal layer** (C) of mostly undifferentiated cells adjacent to the lumen. In several locations, these ependymal cells differentiate into the epithelium covering the choroid plexus, the source of the **cerebrospinal fluid**.

As neuroblasts accumulate adjacent to the ependymal layer, their cell bodies form a **mantle layer** (B) which is the precursor of the grey matter. The **alar plate** (I) is a dorsal concentration of neuronal cell bodies in the mantle layer. It forms sensory neurons. The **basal plate** (J) is a ventral concentration of neuronal cell bodies in the mantle layer. It forms motor neurons. Axons begin to project from these neurons and become myelinated by oligodendroglial cells (in the CNS), forming the **marginal layer** (A) which is the precursor of the white matter.

The mid-dorsal portion of the neural tube remains thin during development. The dorsal **roof plate** (G) perforates to produce 3 foramina (1 foramen of Magendie and 2 foramina of Luschka) for outflow of CSF. In laboratory animals, transplantation experiments have shown that the ventral **floor plate** (H) is crucial for inducing the motor neurons in the adjacent basal plate.

Examine the photograph of a brain from an autopsy in **Figure 3.3** below and then answer the items, choosing the **BEST** response.

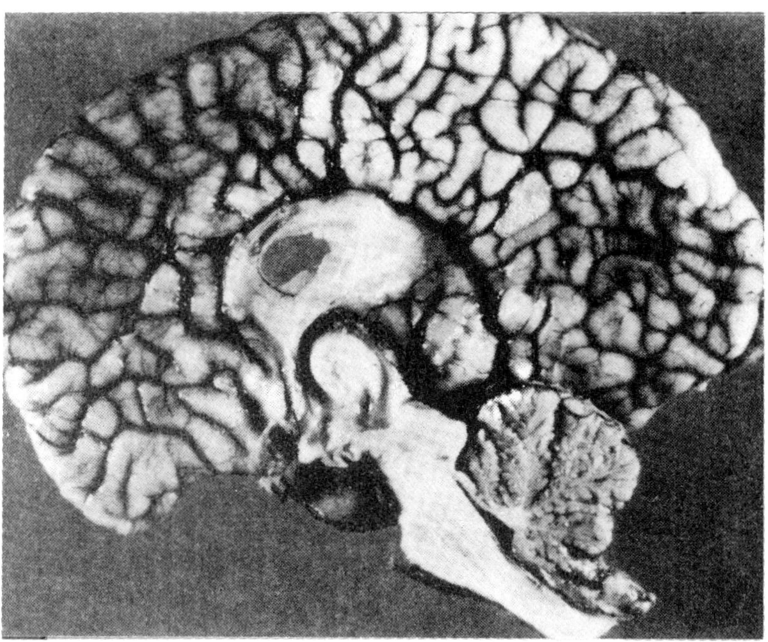

Figure 3.3

142. This brain is an example of which specific clinical entity?

    (A)   hydrocephaly
    (B)   cephalocele
    (C)   Arnold-Chiari malformation
    (D)   anencephaly
    (E)   microcephaly

143. The etiology of this condition is

    (A)   virally induced
    (B)   strictly genetic
    (C)   multifactorial
    (D)   drug induced
    (E)   None of the above

144.   This malformation is **MOST** often associated with

    (A)   hydrocephalus
    (B)   blindness
    (C)   cleft palate
    (D)   conductive hearing loss

145.   This malformation in also associated with

    (A)   spina bifida occulta
    (B)   lumbosacral myelomeningocele
    (C)   rachischisis
    (D)   maternal diabetes mellitus
    (E)   maternal hypertension

ANSWERS AND TUTORIAL ON ITEMS 142-145

The answers are: **142-C; 143-C; 144-A; 145-B**. In many children with **lumbosacral myelomeningocele**, a kind of **spina bifida cystica**, there are associated defects in the brain stem and cerebellum known as the **Arnold-Chiari malformation**. The etiology of this and other neural tube defects is poorly understood but both genetic and environmental factors are suspected, i.e., its etiology is **multifactorial**. In the example shown in **Figure 3.3**, there is elongation of the brain stem and herniation of the lower part of the cerebellum through the foramen magnum. **Hydrocephalus** is frequently associated with the Arnold-Chiari malformation and is most likely secondary to it. One plausible sequence of events follows. Compression of the cerebellum can lead to obstruction of the foramen of Magendie. This leads to intraventricular accumulation of cerebrospinal fluid and thus to hydrocephalus. The Arnold-Chiari malformation is a relatively common CNS abnormality (about 1/1000 births).

A 28-year-old woman of Irish extraction is in the 30th week pregnancy and reports no fetal movements. Maternal serum α-fetoprotein is elevated. Ultrasonography in **Figure 3.4** reveals (FC = fetal cranium; FCS = fetal cervical spine):

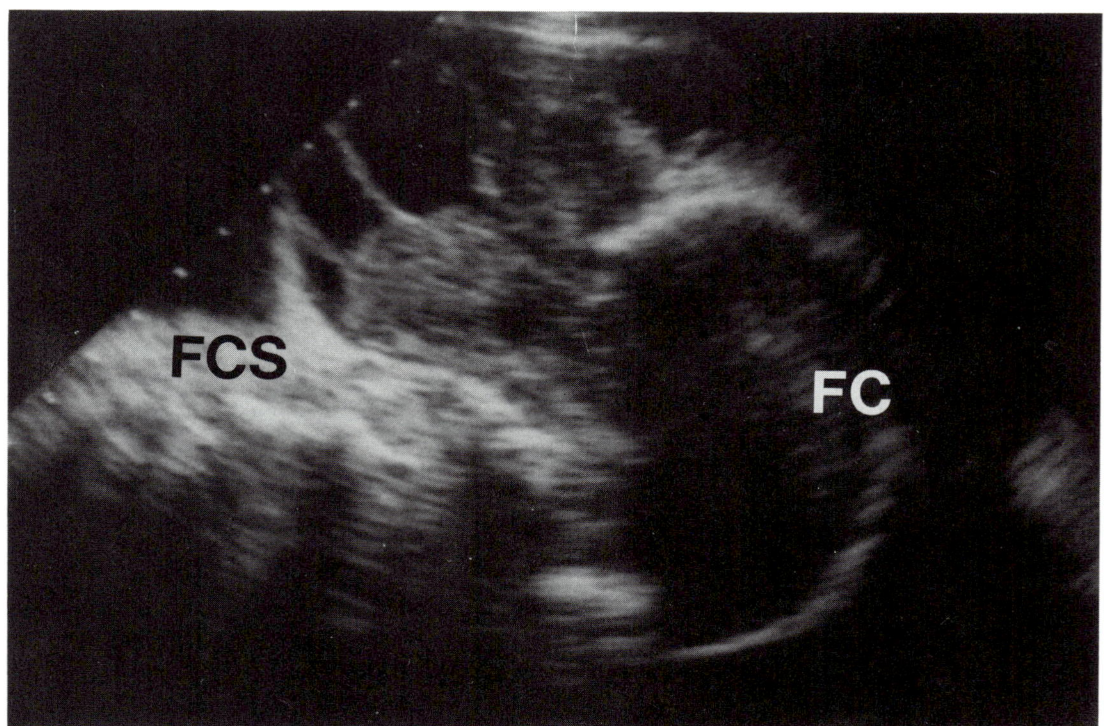

Figure 3.4

146.   The **MOST** accurate diagnosis of this anomaly is

(A)   anencephaly
(B)   spina bifida
(C)   respiratory distress syndrome
(D)   spina bifida occulta
(E)   encephalocele

147.   All of the following statements are true concerning this anomaly **EXCEPT**:

(A)   It is caused by a point mutation.
(B)   The likelihood of recurrence of a similar congenital anomaly in subsequent pregnancies is increased.
(C)   If carried to term, the infant has significant risk of paralysis.
(D)   It is a kind of neural tube defect.
(E)   The fetus can survive after birth.

148. Which of the following represents the **MOST** extensive neural tube defect?

    (A)   spina bifida occulta
    (B)   spina bifida cystica
    (C)   anencephaly
    (D)   myelomeningocele
    (E)   rachischisis

149. The **MOST** likely developmental process causing the congenital anomaly shown in **Figure 3.4** is

    (A)   inadequate neural tube closure
    (B)   excessive production of cerebrospinal fluid
    (C)   excessive skull growth
    (D)   breakage of chromosomes
    (E)   nondisjunction during meiosis

150. The abnormal process causing failure of formation of the skull **MOST** likely is

    (A)   abnormal bone induction
    (B)   excessive production of cerebrospinal fluid
    (C)   inadequate fetal growth
    (D)   defective osteoblasts
    (E)   abnormal placental transport of calcium

ANSWERS AND TUTORIAL ON ITEMS 146-150

The answers are: **146-E; 147-A; 148-E; 149-A; 150-A**. The ultrasonogram in **Figure 3.4** shows an example of a fetus with an **encephalocele**, a type of **neural tube defect**. Neural tube defects show a complex multifactorial etiology with both genetic and environmental factors contributing to an increased incidence. In this case, the base of the occipital bone is poorly formed. During early development, the neural tube induces formation of the bony structures protecting the central nervous system such as the neural arches and the vault of the skull. In many types of neural tube defects, there is both abnormal formation of nervous tissue leading to neurological deficits and abnormal formation of the overlying bony structures that normally protect the nervous system. In encephalocele, meninges protrude into the space created by abnormal formation of the base of the occipital bone. Fetal blood and cerebrospinal fluid contains α-**fetoprotein**. In many neural tube defects, the α-fetoprotein enters the amniotic fluid (where its level can be measured as an indication of neural tube defects) and may cross into the maternal circulation and subsequently be detected by an assay of maternal serum α-fetoprotein.

Neural tube defects range in severity. In **spina bifida occulta**, there is only an absence of the bony neural arch of some vertebrae with no neurological deficits. In **rachischisis**, there is a complete failure of closure of the neural tube, so that the brain and spinal cord do not form. **Anencephalic fetuses** have a failure of closure of the neural tube in the cranial region. Thus, the spinal cord is intact but the brain is malformed.

<u>Items 151-156</u>

Examine the diagram in **Figure 3.5** showing various endocrine organs labeled with a letter. Match the descriptions below of the embryonic development of the labeled organ with the **MOST** appropriate letter. Answers may be used once, more than once, or not at all.

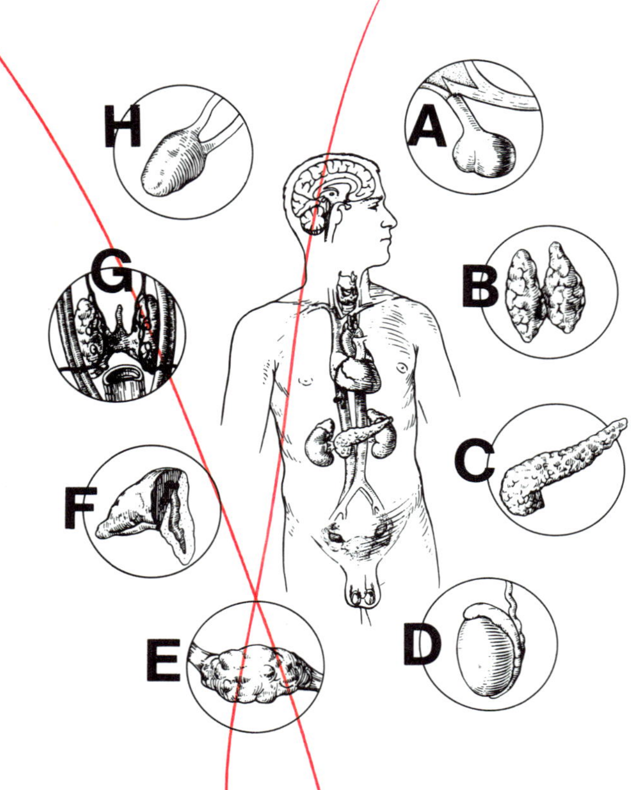

Figure 3.5

151. This organ contains both a neural crest and a coelomic epithelial derivative.

152. This organ is formed from a diverticulum of the foregut. Developmental defects can result in formation of ectopic functional tissue at the root of the tongue. It is not a derivative of the pharyngeal apparatus.

153. This organ is formed from two primordia, one arising in the roof of the stomodeum and the other arising in the floor of the diencephalon.

154.   The endocrine cells of this organ arise from a diverticulum of the cranial portion of
the foregut.

155.   This organ contains functional derivatives from the yolk sac, coelomic epithelium, and
mesonephros. At birth, in the gametogenic portion, all gametogenic cells have 46
chromosomes and the somatic diploid amount of DNA.

156.   Endocrine cells are derived from the third or fourth pharyngeal pouches.

ANSWERS AND TUTORIAL ON ITEMS 151-156

The answers are: **151-F; 152-B; 153-A; 154-C; 155-D; 156-G**. The **pituitary gland** (A) is
derived from two different embryonic primordia. The adenohypophysis (source of major
endocrine secretions of the pituitary, e.g., TSH) is derived from an evagination of the roof of
the stomodeum (ectoderm) called **Rathke's pouch**. Rathke's pouch derivatives include the
pars distalis, pars tuberalis, and pars intermedia of the adenohypophysis. The
neurohypophysis (source of releasing hormones) is derived from an evagination of the floor
of the brain (neuroectoderm) called the **infundibulum**.

The **thyroid gland** (B) arises from a diverticulum of the foregut (endoderm) at the
root of the developing tongue. The primordium of the thyroid gland descends through the
developing neck region as the **thyroglossal duct**. The thyroid gland forms in the most caudal
part of the duct. The cranial part of the duct usually degenerates but may persist, forming
ectopic thyroid tissue along the route of descent. The thyroid follicular epithelial cells, which
secrete thyroxine, are endodermally derived. The calcitonin-secreting parafollicular **C-cells** of
the thyroid gland originate in the **neural crest** (neuroectoderm).

In the **pancreas** (C), both exocrine acinar tissue and endocrine islet tissue, arise from
dorsal and ventral pancreatic diverticula at the duodenal (caudal) portion of the foregut
(endoderm). During gut rotation, the dorsal and ventral pancreatic primordia fuse to form a
single gland. Earlier theories of neural crest origin for islet tissue appear to be unsupported
by experimental evidence.

The **testes** (D) and **ovaries** (E) have important endocrine roles. The chief endocrine
tissue of the testis is a diffuse collection of interstitial **Leydig cells**. These arise from genital
ridge mesenchyme and secrete testosterone. The chief endocrine tissue of the ovary is more
complex. The **granulosa cells** of follicles and the **thecal cells** surrounding follicles both have
an important endocrine role. The thecal cells, like Leydig cells, arise from the genital ridge
mesenchyme. The granulosa cells, like their homolog in the male, the Sertoli cells, arise
from coelomic epithelium. The gametogenic cell lines (oogonia and their products in the
female, spermatogonia and their products in the male) are derived from **primordial germ
cells**, a diploid cell line arising in the yolk sac and migrating to the gonadal primordium
early in development.

The **adrenal glands** (F) have a population of medullary cells that originate in neural
crest (neuroectoderm) and cortical cells that are arise from the coelomic epithelium.

The **parathyroid glands** (G) usually are found as a superior pair (derived from fourth pharyngeal pouch endoderm) and an inferior pair (derived from third pharyngeal pouch endoderm). The position and even number of parathyroid glands can be variable.

The **pineal gland** (H) is derived entirely from the central nervous system. It arises as an evagination of the roof plate of the brain in the region of the epithalamus.

Items 157-158

157. Many congenital birth defects are caused by aneuploidy. Abnormalities in the development of the ocular system are **MOST** severe in which of the following aneuploid conditions?

    (A)   trisomy 21 (Down syndrome)
    (B)   trisomy 13 (Patau syndrome)
    (C)   trisomy 18 (Edwards syndrome)
    (D)   Klinefelter syndrome
    (E)   Turner syndrome

158. Congenital anomalies in the auditory system are commonly found in which of the following conditions?

    (A)   trisomy 18 (Edwards syndrome)
    (B)   mandibulofacial dysostosis (Treacher Collins syndrome)
    (C)   intrauterine rubella virus infection
    (D)   Potter's syndrome
    (E)   All of the above.

ANSWERS AND TUTORIAL ON ITEMS 157 AND 158

The answers are: **157-B; 158-E**. While many different clinical syndromes may include developmental anomalies in the eyes and ears, ocular development is most profoundly disrupted in **Trisomy 13** where microphthalmia is common. In **Trisomy 21**, speckling of the iris and oblique palpebral fissures are common but relatively minor manifestations of the disease.

Different clinical syndromes may affect the development of the auditory system. In **Trisomy 18**, the auricle is dysmorphic in a characteristic way. In mandibulofacial dysostosis, the development of the first pharyngeal arch is altered. Thus, the auricles, external auditory meatus, and middle ear ossicles can be abnormal, leading to conduction hearing loss. **Rubella infections** often target developing neuroepithelium and commonly leading to ocular and auditory abnormalities. **Potter's syndrome** occurs in individuals with bilateral renal agenesis which is usually fatal shortly after birth. It associated with characteristic facial features, oligohydramnios, and auriclar abnormalities.

52

<u>Items 159-163</u>

Examine the photomicrograph of a developing eye in **Figure 3.6** below. Match the labeled structure with the **MOST** appropriate statement concerning its developmental role in the items below. Answers may be used once, more than once, or not at all.

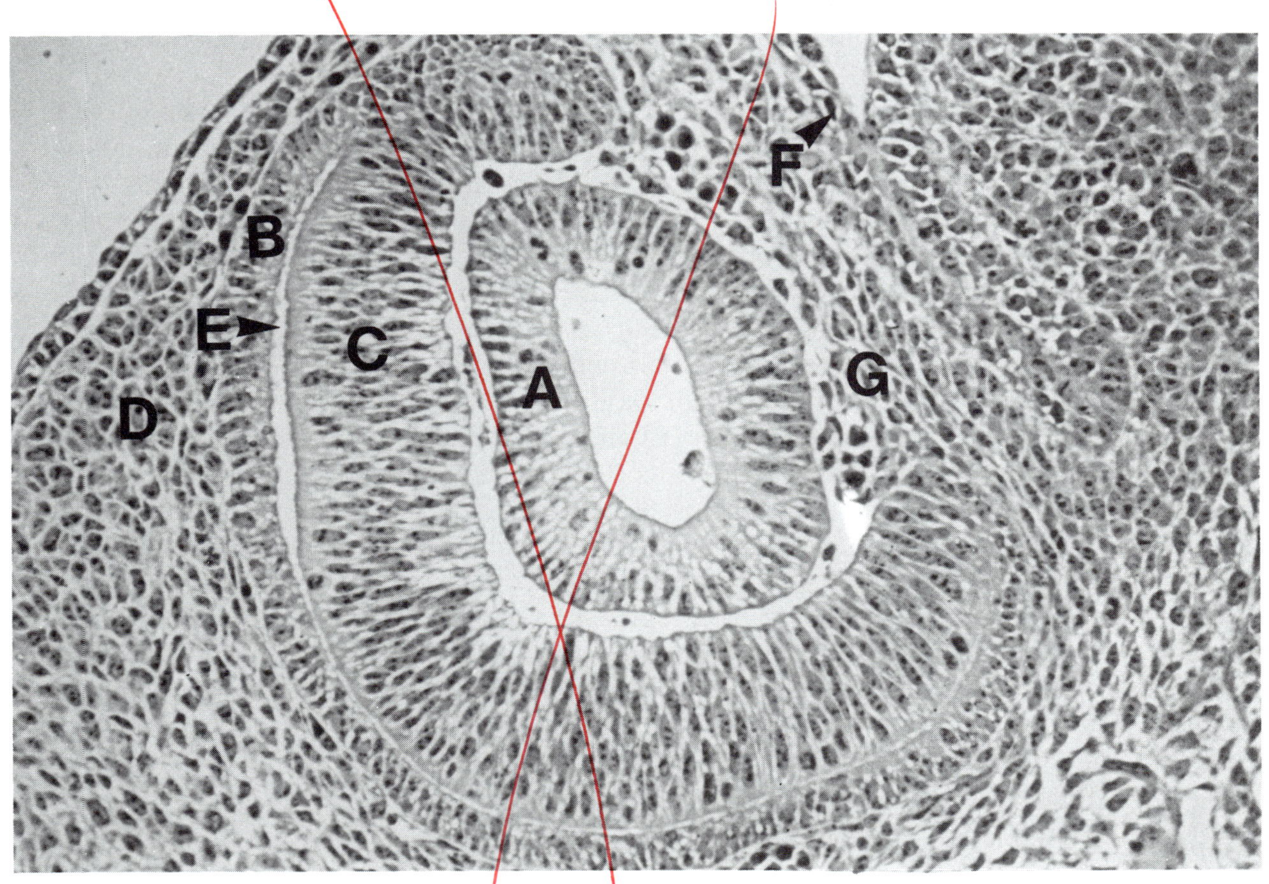

Figure 3.6

159. This structure induces invagination of the optic vesicle.

160. This structure is crucially involved in retinal detachment secondary to head trauma.

161. This structure forms phagocytic cells associated with rod outer segments.

162. This structure forms the corneal epithelium.

163. This structure forms the corneal endothelium and anterior chamber.

The answers are: **159-A; 160-E; 161-B; 162-F; 163-G**. **Figure 3.6** is a histological section of a developing mouse eye. It is approximately equivalent to a developing human eye at about 5 weeks of development. The **lens vesicle** (A) is derived from an invagination of the surface ectoderm. It is induced by the optic vesicle and in turn, induces the invagination of the optic vesicle to form an optic cup. The optic cup has two distinct layers, an outer **pigmented retinal epithelium** (B) and an inner **neural retinal epithelium** (C). The former forms the simple cuboidal epithelium of the definitive pigmented retina. The latter forms the photosensitive portion of the retina, complete with rods, cones, bipolar cells, and ganglion cells which project their axons into the optic stalk, eventually forming the optic nerve. **Posterior head mesenchymal cells** (D) condense around the outer layer of the optic cup and eventually form the choroid layer and the sclera. The **intraretinal space** (E) is a remnant of the lumen of the central nervous system and earlier was in direct communication with the third brain ventricle. It is a potential space in the adult eye where the apical, nonadhesive layers of the pigmented retina and neural retina come in contact. These two layers are loosely bound to each other and can separate (e.g., from a blow to the head) leading to retinal detachment. The **surface ectoderm** (F) anterior to the lens forms the corneal epithelium. **Anterior head mesenchymal cells** (G) contribute to the corneal stroma and corneal endothelium (the name used for this structure in definitive histology textbooks). The corneal endothelium is also called corneal mesothelium, the name used in embryology textbooks. This name does not seem appropriate because mesothelium is usually reserved for the mesodermally derived simple squamous epithelium lining serous body cavities. The anterior chamber of the eye is directly connected to the cardiovascular system via the canal of Schlemm. Therefore, the name corneal endothelium seems preferable. The anterior chamber of the eye also forms from fusion of small vesicles in the mesenchyme posterior to the cornea into a cleftlike space, much like blood vessels, reinforcing the use of the term corneal endothelium.

Examine the labeled drawing of an adult ear in **Figure 3.7** below. Match the statements below on ear development with the **MOST** appropriate labeled structure. Answers may be used once, more than once, or not at all.

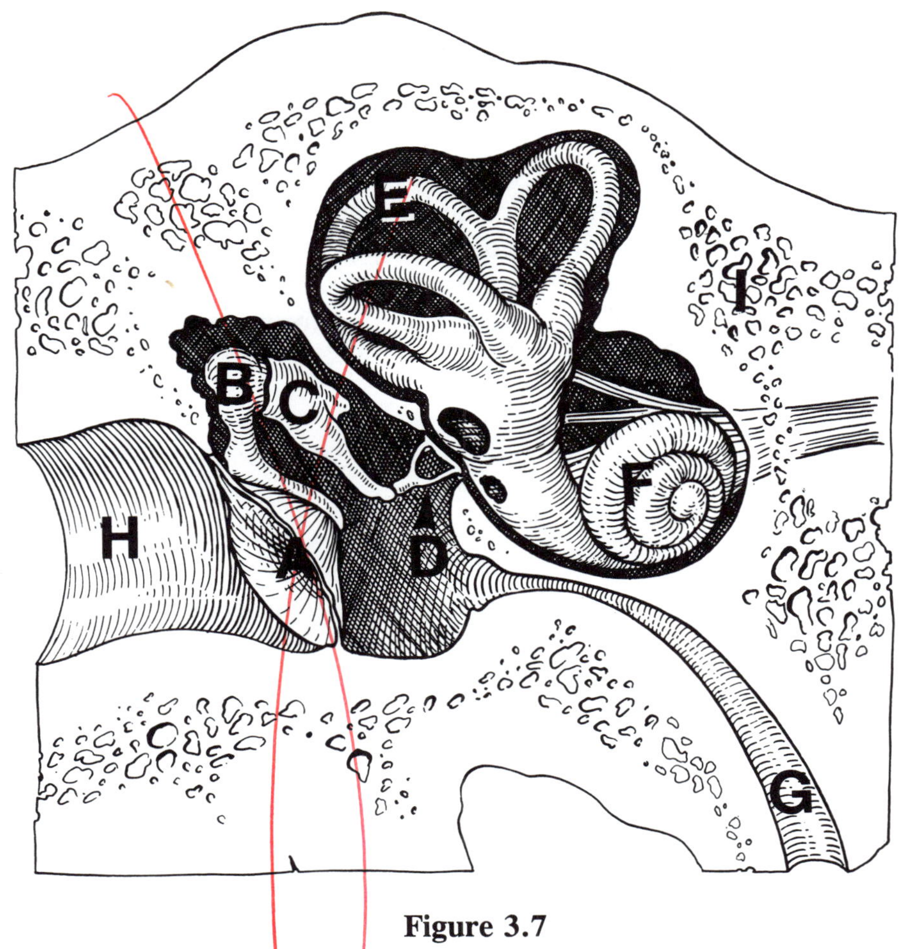

Figure 3.7

164. This structure is formed from the utricular portion of the otocyst.

165. This structure is formed from the saccular portion of the otocyst.

166. This structure is formed from the first pharyngeal pouch.

167. This structure is formed from the first pharyngeal groove.

168. This structure is formed from the otic capsule and will dense bone.

169. This structure is formed from second pharyngeal arch mesenchyme.

The answers are: **164-E; 165-F; 166-G; 167-H; 168-I; 169-D**. The **tympanic membrane** (A) is a derivative of the first pharyngeal membrane which lies between the first pharyngeal groove [**external auditory meatus** (H) precursor] and the first pharyngeal pouch [**auditory** or eustachian **tube** (G) precursor].

The **malleus** (B), **incus** (C) and **stapes** (D) are the three middle ear ossicles. They conduct vibrations from the tympanic membrane to the oval window. The malleus and incus form in the first pharyngeal arch. The stapes forms in the second pharyngeal arch.

The **bony labyrinth** (I) is a cavity within the petrous temporal bone. It contains the membranous labyrinth which is derived from the **otocyst**. The otocyst is formed by invagination of surface ectoderm and undergoes a shape change to form a **utricular portion** and a **saccular portion**. The utricular portion forms the **semicircular ducts** (E), endolymphatic duct, and the utricle. The saccular portion forms the saccule and the **cochlear duct** (F).

Items 170-171

170.   In what month of pregnancy does spontaneous neuromuscular movement begin in the embryo/fetus?

    (A)   second
    (B)   third
    (C)   fourth
    (D)   fifth
    (E)   sixth

171.   In which developmental period do reflex movements begin?

    (A)   first two weeks
    (B)   embryonic
    (C)   second trimester
    (D)   third trimester
    (E)   neonatal

ANSWERS AND TUTORIAL ON ITEMS 170-171

The answers are: **170-A; 171-A**. **Spontaneous neuromuscular movement** has been observed in late embryos. Skeletal muscles become functional as soon as they appear. **Reflex movements** have been elicited near the end of the embryonic period by light stroking of the perioral region. Initially, there is a whole body response caused by the contraction of groups of back muscles. By the end of the first trimester light stroking of the skin produces discrete local movements.

CHAPTER IV
HEMATOPOIESIS AND IMMUNOLOGIC DEVELOPMENT

Items 172-177

Items 172 through 177 pertain to hematopoiesis. Choose the **BEST** response.

172. Which cell type is formed earliest in development?

   (A)   red blood cell
   (B)   B lymphocyte
   (C)   T lymphocyte
   (D)   neutrophil
   (E)   monocyte

173. Which location in the body is the earliest major hematopoietic site?

   (A)   allantois
   (B)   yolk sac
   (C)   spleen
   (D)   liver
   (E)   bone marrow

174. Prior to bone marrow formation, which structure is the **MOST** important site of hematopoiesis?

   (A)   dermis
   (B)   yolk sac
   (C)   spleen
   (D)   liver
   (E)   thymus

175. All of the following organs contain erythropoietic stem cells during development **EXCEPT**:

   (A)   yolk sac
   (B)   spleen
   (C)   liver
   (D)   bone marrow
   (E)   thymus

176. The cytotoxic lymphocytes that protect us from cells which have undergone malignant transformation are programmed to differentiate after passage through which organ?

    (A)   yolk sac
    (B)   spleen
    (C)   liver
    (D)   bone marrow
    (E)   thymus

177. Which of the following organs is capable of hematopoiesis postnatally under pathologic conditions?

    (A)   red marrow
    (B)   yellow marrow
    (C)   liver
    (D)   spleen
    (E)   all of the above

ANSWERS AND TUTORIAL ON ITEMS 172-177

The answers are: **172-A; 173-B; 174-D; 175-E; 176-E; 177-E**. **Hematopoiesis** begins very early in development. Primitive red blood cells are first detectable in the embryo as early the end of the second week of development, and soon after this, there are numerous anastomotic vessels formed in the **yolk sac**, connecting stalk, and chorionic villi. These vessels contain primitive (nucleated) red blood cells. By the middle of the fourth week, an unsepted but beating heart tube propels nucleated red blood cells around the embryonic body, through the connecting stalk into the chorionic villi.

Relatively minor erythropoiesis and still less myelopoiesis occurs in the **spleen** during the third to fifth months. The spleen has a relatively minor hematopoietic role. Beginning in the sixth week, the **liver** is the major hematopoietic organ in the body, but by the second month, the **bone marrow** gradually becomes dominant. Bone marrow, first present in the clavicles and then in other bones as they develop, takes on an increasingly important hematopoietic role as the hematopoietic role of the liver declines. Postnatally, hematopoiesis is restricted to the red bone marrow but the potential for **extramedullary hematopoiesis** remains in the liver and even in the spleen. In cases of severe hemolytic anemia or with failure of bone marrow, all of the organs listed in Item 177, even the spleen, can become reactivated to produce blood cells.

Erythropoietic stem cells are found in all of the organs listed in Item 176 except for the thymus. Lymphopoiesis and cytotoxic T lymphocyte differentiation occurs in the thymus but the undifferentiated T lymphoblasts arise in the yolk sac and liver and move chemotactically to the thymus (under the influence of thymotactin) where they are programmed to differentiate into fully functional T lymphocytes.

<u>**Items 178-182**</u>

The following set of items pertains to basic embryology of immune system organs. Choose the **BEST** response.

178.   Which of the following immune organs is formed by rudiments derived from **BOTH** mesoderm and endoderm?

    (A)   palatine tonsil
    (B)   spleen
    (C)   axillary lymph nodes
    (D)   Peyer's patches
    (E)   inguinal lymph nodes

179.   The epithelium covering cryptic infoldings in the palatine tonsil is derived from the

    (A)   first pharyngeal pouch
    (B)   first pharyngeal groove
    (C)   second pharyngeal pouch
    (D)   second pharyngeal groove
    (E)   third pharyngeal pouch
    (F)   third pharyngeal groove

180.   The majority of lymphocytes in the palatine tonsil in a 5-year-old child with bacterial tonsillitis were derived from precursors in the

    (A)   liver
    (B)   yolk sac and liver
    (C)   spleen
    (D)   thymus
    (E)   none of the above

181.   All of the following statements concerning development of the spleen are true **EXCEPT**:

    (A)   Splenic endothelial cells in the red pulp are derived from endoderm.
    (B)   Splenic lymphocytes are derived from the yolk sac and liver.
    (C)   Many of the cells in the periarterial lymphatic sheath are processed in the thymus.
    (D)   During the first trimester the spleen is erythropoietic and myelopoietic.
    (E)   The spleen forms in the dorsal mesogastrium.

182.   Thymic reticular epithelial cells are derived from

   (A)   second pharyngeal pouch only
   (B)   first pharyngeal pouch only
   (C)   fourth pharyngeal pouch only
   (D)   the same rudiment as the principal cells of the inferior parathyroid glands
   (E)   none of the above

ANSWERS AND TUTORIAL ON ITEMS 178-182

The answers are: **178-A; 179-C; 180-B; 181-A; 182-D**. The surface of the **palatine tonsil** contains cryptic infoldings of the pharyngeal mucosa. The crypts are lined with stratified squamous epithelium that is derived from the **second pharyngeal pouch**. The adjacent lamina propria becomes seeded with B and T cells derived from the bone marrow. The T cells are programmed to differentiate after passage through the thymus.

The **spleen** arises in the **dorsal mesogastrium** (dorsal mesentery of the stomach). The mesenchymal cells in the spleen become arranged into vascular channels, eventually leading to the formation of red pulp. The white pulp arises as lymphocytes become seeded into the organ from the bone marrow. Splenic T cells, which are the predominant cell type in the periarterial lymphatic sheaths and are also present in the marginal zone, are derived from the yolk sac and liver but are processed to differentiate into T cells after passage through the thymus. B cells are concentrated in the lymphatic nodules of the white pulp and are found in abundance in the marginal zone.

The **thymus** has a reticular stroma of reticular epithelial cells. These cells are derived from the **third pharyngeal pouch**, along with the inferior parathyroid glands. The precursors of thymic lymphocytes originate in the yolk sac and liver. The thymus is the major organ for programming differentiation of pre-T lymphoblasts. It is also extremely important as a lymphopoietic organ. Pre-T lymphoblasts migrate to the thymus. After complex interaction with thymic reticular epithelial cells, mediated by several "thymic hormones" including several kinds of **interleukins** (IL-1 and IL-4), a cascade of events leads to T cell differentiation. These T cells are then gradually seeded to widely scattered lymph nodules in for example, the enteric mucosa and submucosa, lymph nodes, and the spleen.

The following set of items pertains to development of immune competence in an infant.
Choose the **BEST** response.

183.   In a new born infant the majority of the circulating IgG molecules is derived from

(A)   fetal B cells
(B)   fetal plasma cells
(C)   maternal T cells
(D)   fetal T cells
(E)   passive immunization via placental transport of maternal IgGs

184.   Breast-fed babies often have fewer gastrointestinal infections during early postnatal life. Which kind of immunoglobulin in colostrum is **MOST** closely associated with this function?

(A)   IgA
(B)   IgD
(C)   IgE
(D)   IgG
(E)   IgM

185.   Which component of secretory immunoglobulins renders them less susceptible to enteric proteolysis?

(A)   H chain
(B)   L chain
(C)   J polypeptide
(D)   secretory component
(E)   variable region

186.   The heavy (H) and light (L) chains of secretory immunoglobulins in colostrum are synthesized

(A)   by plasma cells deep to the luminal epithelium of the mammary gland
(B)   in the infant's oral cavity
(C)   in the infant's salivary glands
(D)   in ductal epithelial cells of the mammary gland
(E)   in acinar epithelial cells of the mammary gland

187.   Which secretions contain secretory immunoglobulins?

(A)   milk
(B)   saliva
(C)   sweat
(D)   tears
(E)   all of the above

ANSWERS AND TUTORIAL ON ITEMS 183-187

The answers are: **183-E; 184-A; 185-D; 186-A; 187-E**. If there is no prenatal intrauterine infection, the fetus normally develops in a sterile environment and therefore is not exposed to bacterial antigens. The mother mounts an immune response to bacterial antigens that she has been exposed to (and to which the new born infant is likely to be exposed). Maternal IgGs are then transported across the placenta, leading to **passive immunization** of the fetus. These maternally derived IgGs are slowly degraded after birth and are replaced by new IgGs produced by the infant's own immune system, secondary to exposure to environmental antigens.

Human breast milk is also an important potential source of immune protection for an infant. Breast milk is rich is several nonspecific antibacterial substances. It also has an abundant supply of immunoglobulins including **secretory IgA**. The H and L chains of this IgA are synthesized in plasma cells in the connective tissue of the mammary gland. IgA molecules are then joined by a **J-polypeptide** and bound to the epithelial cells in acini where a **secretory component** is added. The secretory component covers the protease-sensitive regions of the IgA molecules. These **proteolysis-resistant molecules** are then ingested by the infant where they provide significant protection from enteric infections. Secretory IgAs are abundant in other bodily fluids also including sweat, tears, saliva, and vaginal secretions.

Items 188-191

You are a neonatologist on-call on a busy obstetrical service. You examine a newborn male infant with tetany, cyanosis, and unusual craniofacial features including mandibular hypoplasia, a reduced philtrum, and low-set ears. Laboratory findings reveal hypocalcemia, decreased parathormone levels, but a nearly normal immunoglobulin profile.

188. The **MOST** likely diagnosis of this patient's condition is

    (A)   Down syndrome
    (B)   DiGeorge syndrome
    (C)   Bruton's agammaglobulinemia
    (D)   mandibulofacial dysostosis
    (E)   fetal alcohol syndrome and AIDS

189. All of the following would also commonly be found in patients with the condition **EXCEPT**:

    (A)   *Pneumocystis* infections and oral candidiasis
    (B)   reduced number of B cells
    (C)   increased number of T cells
    (D)   atrial septal defects
    (E)   hyperphosphatemia

190. Agenesis of which embryonic structures usually produces this condition?

    (A)   thyroglossal duct
    (B)   second pharyngeal arch mesenchyme
    (C)   second pharyngeal pouch
    (D)   third and fourth pharyngeal pouches
    (E)   septum transversum

191. In many cases of this syndrome, minor deletions have been detected in the short arm of which chromosome?

    (A)   20
    (B)   22
    (C)   X
    (D)   13
    (E)   Y

ANSWERS AND TUTORIAL ON ITEMS 188-191

The answers are: **188-B; 189-C; 190-D; 191-B**. The patient is afflicted by **DiGeorge syndrome**, an immunodeficiency syndrome caused by **thymic hypoplasia**. The incidence of DiGeorge syndrome is approximately 1 in 20,000 with a M:F ratio of 1:1. Deletions in the short arm of chromosome 22 have been detected in the vast majority of cases. The severity of this syndrome is highly variable, depending upon the degree of organ hypoplasia. The immunologic problems in these infants (e.g., recurrent infections by opportunistic organisms) are due to thymic aplasia or hypoplasia leading to complete or partial absence of cell-mediated immune responsiveness. The hypocalcemia is due to variable **parathyroid hypoplasia**. Patients with DiGeorge syndrome are commonly afflicted with atrial or ventricular septal defects and craniofacial anomalies. In DiGeorge syndrome, the thymic and parathyroid hypoplasia is due to failed development of the **third and fourth pharyngeal pouches**. Abnormal neural crest cell formation may account for the characteristic facies and cardiovascular defects associated with DiGeorge syndrome.

Items 192-197

In a second pregnancy, a 25-year-old woman delivered a hydropic infant with severe edema, jaundice, and hepatosplenomegaly. Her first child was delivered at home by her aunt without any medical intervention or complications. Her blood type is O negative and her husband's blood type is O positive.

192. The blood type of the infant is

    (A)   A positive
    (B)   A negative
    (C)   O positive
    (D)   O negative
    (E)   B positive

193. All of the following statement concerning the fetus' erythrocytes are true **EXCEPT**:

    (A)   They have the Rh antigen.
    (B)   Their destruction of erythrocytes is minimal.
    (C)   They are produced in the liver.
    (D)   They are produced in the bone marrow.
    (E)   They are coated with maternal anti-Rh IgG.

194. The **MOST** appropriate description of the cause of fetal hepatosplenomegaly is

    (A)   maternal IgGs stimulate hepatic parenchymal cell proliferation
    (B)   extramedullary hematopoiesis
    (C)   increased conjugation of heme degradation products
    (D)   increase in fetal blood volume
    (E)   compensation for reduced placental function

195. The primary cause of Rh isoimmunization is

    (A)   abnormal hepatic morphogenesis
    (B)   mixing of fetal and maternal blood
    (C)   accumulation of maternal IgGs in the placenta
    (D)   failure of transplacental transport of maternal IgGs
    (E)   maternal formation of anti-spermatozoa IgGs

196. If the jaundice in the infant did not subside, which of the following symptoms would **MOST** likely become of major concern?

    (A)   renal failure
    (B)   bowel ischemia
    (C)   brain damage
    (D)   respiratory distress
    (E)   tachypnea

197. Which drug routinely administered postnatally in cases of Rh isoimmunization is likely to ameliorate problems in subsequent pregnancies?

    (A)   ibuprophen
    (B)   prednisone
    (C)   aspirin
    (D)   Rhogam
    (E)   phenobarbital

ANSWERS AND TUTORIAL ON ITEMS 192–197

The answers are: **192-C; 193-B; 194-B; 195-B; 196-C; 197-D**. **Rh isoimmunization** is common is fetuses of **Rh- mothers** and Rh+ fathers. The Rh antigen is controlled by a dominant gene and therefore the fetus of this couple would be Rh+. Rh isoimmunization becomes more severe with each pregnancy because the maternal immune system has a memory for Rh antigen. Normally, the placental trophoblast provides a functional barrier between fetal and maternal blood, although **microvascular leakage** is inevitable. When fetal Rh+ RBCs enter the circulation of an Rh- mother, they are recognized as foreign by the mother and a strong immune response is produced. The resulting maternal IgGs are transported across the placenta where they coat fetal RBCs and result in their rapid destruction in the liver and spleen. The resulting anemia triggers splenic and hepatic (extramedullary) and bone marrow erythropoiesis and results in **hepatosplenomegaly**. Excessive destruction of fetal RBCs leads to the accumulation of heme breakdown products such as bilirubin that causes **jaundice**. Neonatal jaundice can be severe enough to result in brain damage. Administration of **Rhogam** postnatally clears the anti-Rh antibodies from the maternal circulation, ameliorating Rh isoimmunization in subsequent pregnancies.

Items 198-200

An apparently normal male infant began to develop severe, multiple infections at 3 months after birth. Conjunctivitis, otitis media, and cutaneous bacterial infections were noted. The CBC was nearly normal but plasma cells were entirely absent. T cells were present.

198.   This child suffers from

    (A)   Bruton's agammaglobulinemia
    (B)   DiGeorge syndrome
    (C)   adenosine deaminase deficiency
    (D)   malnutrition
    (E)   AIDS

199.   All of the following tissues would be severely hypoplastic **EXCEPT**:

    (A)   palatine tonsils
    (B)   thymus
    (C)   enteric mucosal lymph nodules
    (D)   axillary lymph nodes
    (E)   adenoids

200.   The male:female ratio for occurrence of this disease is

    (A)   1:0
    (B)   1:1
    (C)   1:4
    (D)   1:10
    (E)   1:100

68

ANSWERS AND TUTORIAL ON ITEMS 198-200

The answers are: **198-A; 199-B; 200-A**. This child suffers from X-linked infantile agammaglobulinemia (**Bruton's agammaglobulinemia**). This immunodeficiency disease is rare (1 in 50,000 live births) and is caused by a failure of **pre-B cells** to differentiate into B cells. The disease is restricted to males and is characterized by onset of recurrent respiratory tract infections and cutaneous infections beginning after passively obtained maternal IgGs are depleted (3-6 months after birth).

Due to the lack of B cell differentiation, there is an absence or severe hypoplasia of lymphoid tissue without either germinal centers or plasma cells. There is no **humoral immune response** and thus great susceptibility to bacterial infections beginning soon after birth. The thymus is essentially normal with normal T cells present.

Items 201-205

Match the hematopoietic stem cell in the list below with the **MOST** appropriate description of its developmental history in the answers below. Answers may be used once, more than once, or not at all.

     (A)   PHSC
     (B)   CFU-B
     (C)   CFU-E
     (D)   CFU-GM
     (E)   CFU-Eo
     (F)   CFU-Me
     (G)   CFU-Ma
     (H)   LSC

201.   Ultimately destined to form only platelets.

202.   Can form all cells of the granulocyte, agranulocyte, and erythroid series.

203.   Can form eosinophils but nothing else.

204.   Can form both neutrophils and monocytes but not erythrocytes.

205.   Can form only lymphocytes.

ANSWERS AND TUTORIAL ON ITEMS 201-205

The answers are: **201-F; 202-A; 203-E; 204-D; 205-H**. All of the formed elements of the blood arise from a common **pluripotent hematopoietic stem cell** (PHSC) (A) found in the bone marrow of fetuses and postnatally. When a PHSC divides, it forms a pair of cells, one which remains as a proliferative stem cell and the other becomes a differentiating PHSC. This cell then becomes committed to forming a **lymphocyte stem cell** (LSC) (H) which gives rise to the lymphocytes and plasma cells or another committed cell types. These include:

1) the **erythroid colony-forming unit** (CFU-E) (C) or **erythroid burst-forming unit** (CFU-B) (B), both of which now differentiate into proerythroblasts and subsequent cells of the erythroid series, eventually leading to the formation of erythrocytes.

2) The **CFU-GM** (D) is a progenitor for either neutrophils (via the CFU-G) or monocytes and macrophages via the CFU-M.

3) The **CFU-Eo** (E) is a progenitor for eosinophils. There is also a separate CFU dedicated to basophil production.

4) The **CFU-Me** (F) and **CFU-Ma** (G) are progenitors for megakaryocytes (platelet precursors) and mast cells respectively.

The first distinct precursors of granulocytes are **myeloblasts** which become **promyelocytes** when they acquire a population of nonspecific **azurophilic granules**. Next, they differentiate into myelocytes with the acquisition of their specific neutrophilic, eosinophilic or basophilic granules. Following nuclear condensation and lobulation, these cells are released into the systemic circulation.

CHAPTER V
DEVELOPMENT OF THE HEAD AND NECK

A six-month-old female infant is referred to you because of an abnormality in her upper lip and roof of her oral cavity. Upon examination you observe the defect shown in **Figure 5.1**.

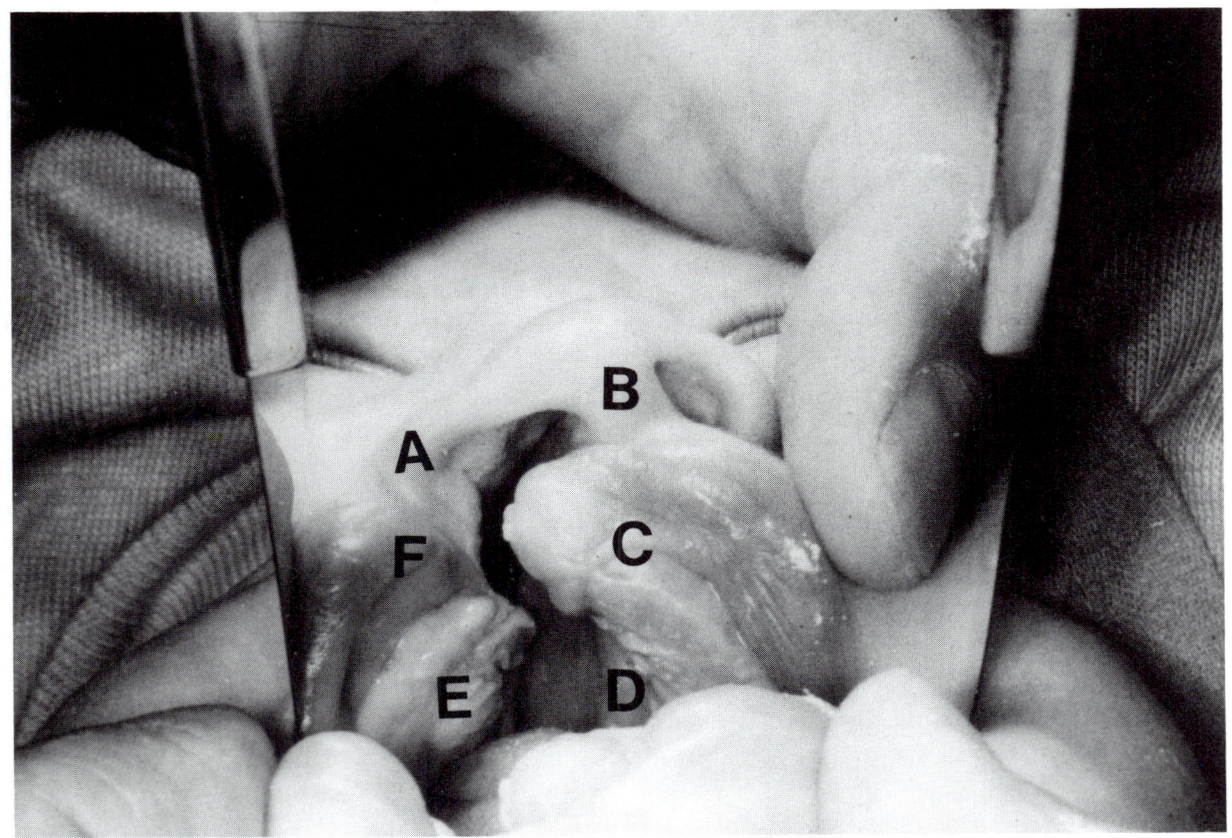

Figure 5.1

Choose the **BEST** response.

206. Which of the following is a **TRUE** statement about clefts of the lip and palate?

(A) Most cases of cleft lip and palate are of multifactorial inheritance and not associated with other anomalies.
(B) Cleft lip and palate can be prevented by folic acid supplementation during pregnancy.
(C) Most children with cleft lip and palate will have below normal intelligence.
(D) Cleft palate results from a breakdown of the previously formed hard and soft palate junction.
(E) Repair of cleft lip and palate is difficult and usually not performed until a child is two years of age.

207. Clefts of the lip and palate

(A) are uncommon
(B) are classified according to their relationship to the nasopalatine foramen
(C) are classified as posterior clefts when they include the alveolar part of the maxilla
(D) occur together more commonly in males than females
(E) form when mesenchymal masses on each side of the cleft fail to fuse

208. In **Figure 5.1**, the tissue at C contains the incisor tooth buds and forms the median part of the upper lip. This region arises from which structure in the six week embryo?

(A) lateral nasal prominence
(B) medial nasal prominence
(C) intermaxillary segment
(D) lateral palatine process
(E) nasal septum

209. The lateral palatine processes are indicated at

(A) A and B
(B) C and F
(C) C and D
(D) B and C
(E) E and D

38-41

ANSWERS AND TUTORIAL ON ITEMS 206-209

38 39 40 41

The answers are: **206-A; 207-E; 208-C; 209-E**. Most cases of **cleft lip/palate** are considered to be of **multifactorial inheritance** and are caused by the failure of fusion of the mesenchyme on each side of the cleft. Surgical repair of clefts of the palate is often delayed until cranial growth is reduced, the abnormal communication between the oral and nasal cavities often being closed initially by a prosthesis. On the other hand, clefts of the lip are repaired as soon as possible after birth. Clefts of the lip/palate are the most common malformation in man and are classified according to their relation to the **incisive foramen**; those posterior to the foramen are midline while those anterior are oblique. Most children with this anomaly have normal intelligence. Cleft lip occurs more often in males while cleft palate with or without cleft lip occurs more often in females.

The **intermaxillary segment** (C) is formed by the lower part of the medial nasal prominences and gives rise to the median part of the upper lip, the incisive tooth buds and a small part of the palate anterior to the incisive foramen. In the embryo this region forms the primary palate. The **lateral palatine processes** (D and E) or shelves begin to fuse with each other and the **nasal septum** (B) during the latter part of the second month but fusion is not complete until near the end of the third month.

A six-year-old Caucasian male was brought to the clinic and presented with the defect shown in **Figure 5.2**.

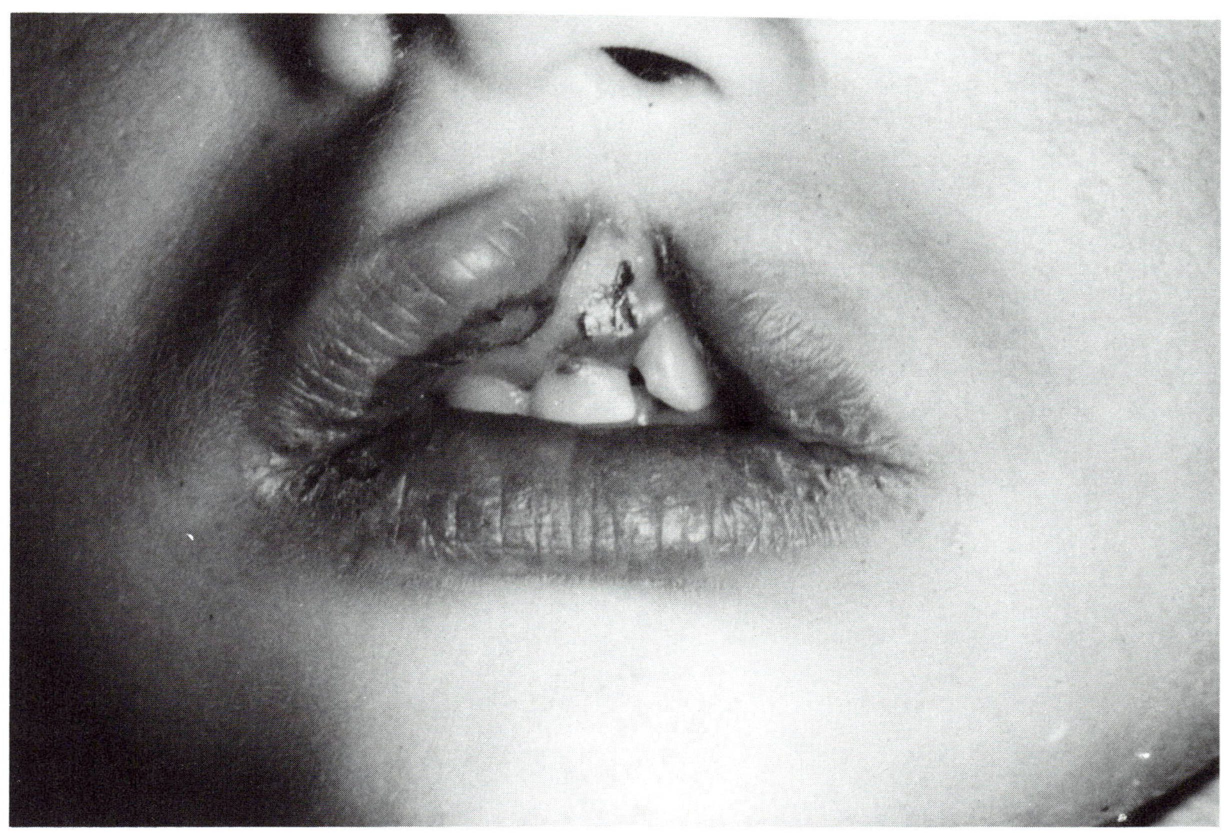

Figure 5.2

210. This cleft is **BEST** classified as

    (A)   bilateral, partial, lateral
    (B)   unilateral, complete, lateral
    (C)   partial, median
    (D)   unilateral, partial, lateral
    (E)   unilateral, complete, median

211. The cleft lip shown in **Figure 5.2** results from which of the following?

    (A)   persistence of a portion of the nasolabial groove
    (B)   partial fusion of the medial nasal prominences
    (C)   failure of complete fusion of the medial nasal and maxillary prominences
    (D)   reduced proliferation of mesenchyme in the lateral nasal prominence
    (E)   persistence of a portion of the nasal placode

ANSWERS AND TUTORIAL ON ITEMS 210 AND 211

The answers are: **210-D; 211-C**. The cleft shown in **Figure 5.2** is only on one side (unilateral), does not extend into the naris (partial) and is lateral to the midline. This kind of cleft forms when the groove between the medial nasal and maxillary prominences deepens rather than filling out by underlying mesenchymal proliferation.

Examine the child shown in **Figure 5.3**. Then choose the **BEST** response to the items below.

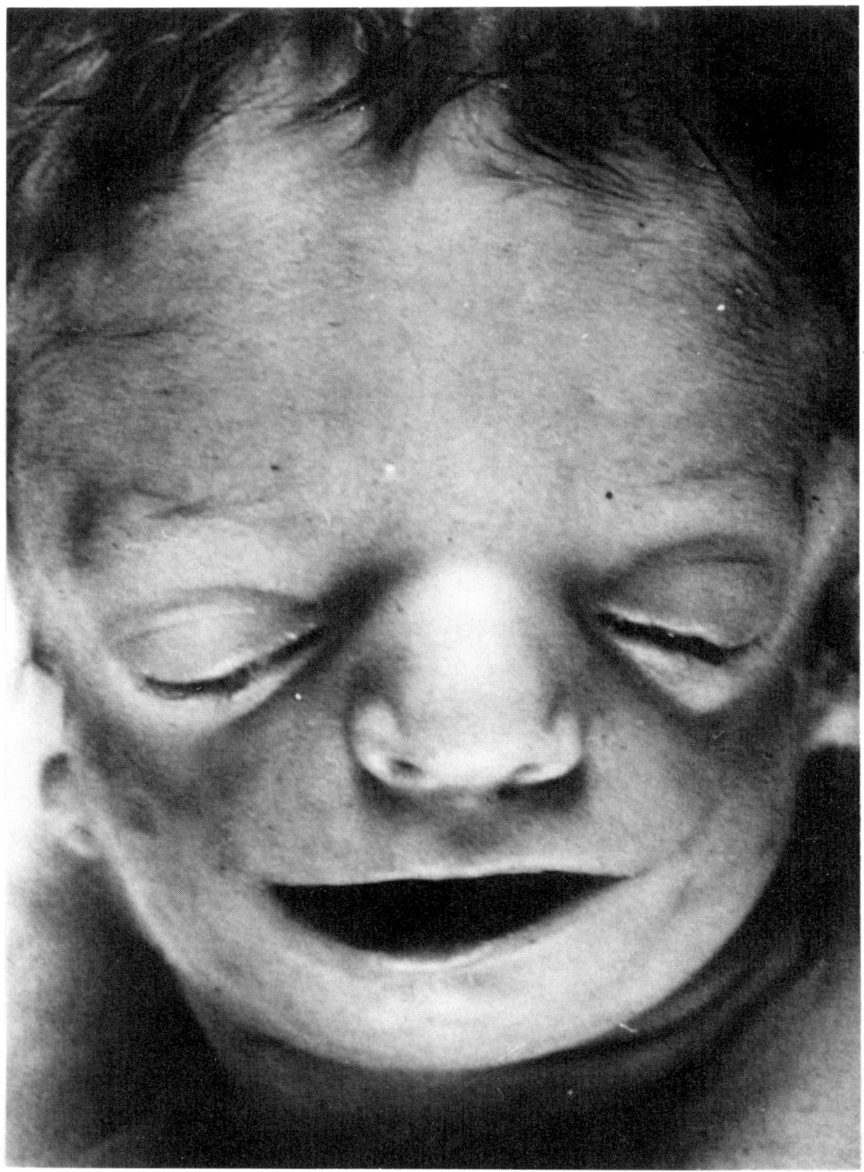

Figure 5.3

212. This child is **MOST** likely afflicted with which of the following conditions?

    (A)   fetal alcohol syndrome
    (B)   trisomy 21
    (C)   DiGeorge syndrome
    (D)   Treacher Collins syndrome
    (E)   trisomy 18

213. All of the following statements concerning this condition are correct **EXCEPT**:

    (A)   probably caused by one or more autosomal dominant genes
    (B)   often associated with deformities of the ears
    (C)   often associated with down slanting palpebral fissures
    (D)   often associated with poorly developed zygomatic bones
    (E)   condition precludes a normal life span

214. Individuals with this condition

    (A)   exhibit maldevelopment of mainly second pharyngeal arch components.
    (B)   rarely exhibit irregular tooth eruption even through the mouth is often large.
    (C)   often exhibit hypoplasia of the hyoid apparatus resulting in swallowing difficulties.
    (D)   typically exhibit mandibulofacial dysostosis.
    (E)   probably produced supernumerary neural crest cells during the embryonic period.

ANSWERS AND TUTORIAL ON ITEMS 212-214

The answers are: **212-D; 213-E; 214-D**. Individuals with **mandibulofacial dysostosis** (Treacher Collins syndrome, first arch syndrome) have a normal life span and normal mental capacity. All of the other choices are correct. This condition is a manifestation of first arch dysgenesis accompanied by maxillary and mandibular hypoplasia and malocclusion from irregular tooth eruption. The syndrome is thought to be caused by insufficient production of neural crest cells.

The child pictured in **Figure 5.4** has an anomaly in the development of the face. Match each of the embryonic precursors below with the **MOST** appropriate labeled structure in the child's face. Labaeled structures may be used once, more than once, or not at all.

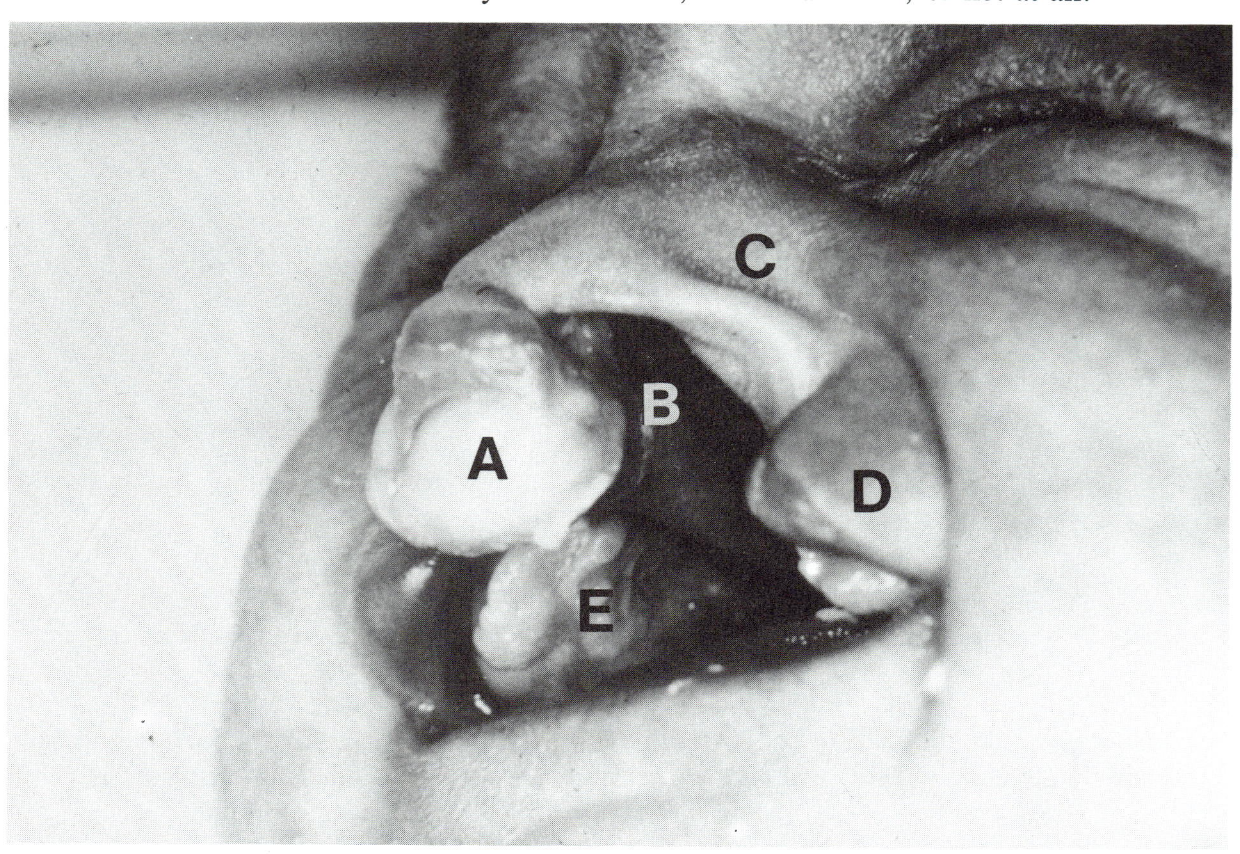

Figure 5.4

215. Maxillary prominence

216. Lateral nasal prominence

217. Medial nasal prominence

218. Tissue between nasal pits

219. Lateral palatine process or shelf

78

Choose the **BEST** response.

220. Which phrase **MOST** accurately describes this condition?

(A)   bilateral cleft of lip and palate
(B)   unilateral cleft of lip
(C)   bilateral cleft of palate
(D)   unilateral cleft of palate
(E)   unilateral cleft of lip and palate

221. The nasal placode

(A)   is a thickening of endoderm
(B)   is located initially in the floor of the stomodeum
(C)   gives rise to olfactory epithelium
(D)   gives rise to the mucosa on the inferior nasal concha
(E)   is a midline structure

ANSWERS AND TUTORIAL ON ITEMS 215-221

The answers are: **215-D; 216-C; 217-A; 218-B; 219-E; 220-A; 221-C**. The lateral part of the upper lip develops from the **maxillary prominence** (D). The side of the nose forms from the **lateral nasal prominence** (C). The lower part of the medial nasal prominences together constitute the **intermaxillary segment** (A) that forms the median part of the upper lip and other structures. This part of the upper lip failed to develop adequately in the patient shown presenting instead an island of tissue in the midline in front of the nasal septum. The **nasal septum** (B) forms from the tissue located between the nasal pits (B) in the embryo. On the right side, the part of the **palate** (E) located in the roof of the oral cavity is formed by the **lateral palatine process** (E). This patient has a bilateral cleft of the lip and palate. The **nasal placode** is a thickening of the ectoderm in the floor of the nasal pit and gives rise to the **olfactory epithelium** that is located in the superior part of the definitive nasal cavity.

Items 222-227

The posterior part of the right auricle of the child shown in **Figure 5.5** is deformed. The hearing ability of the child also is reduced significantly on the right side.

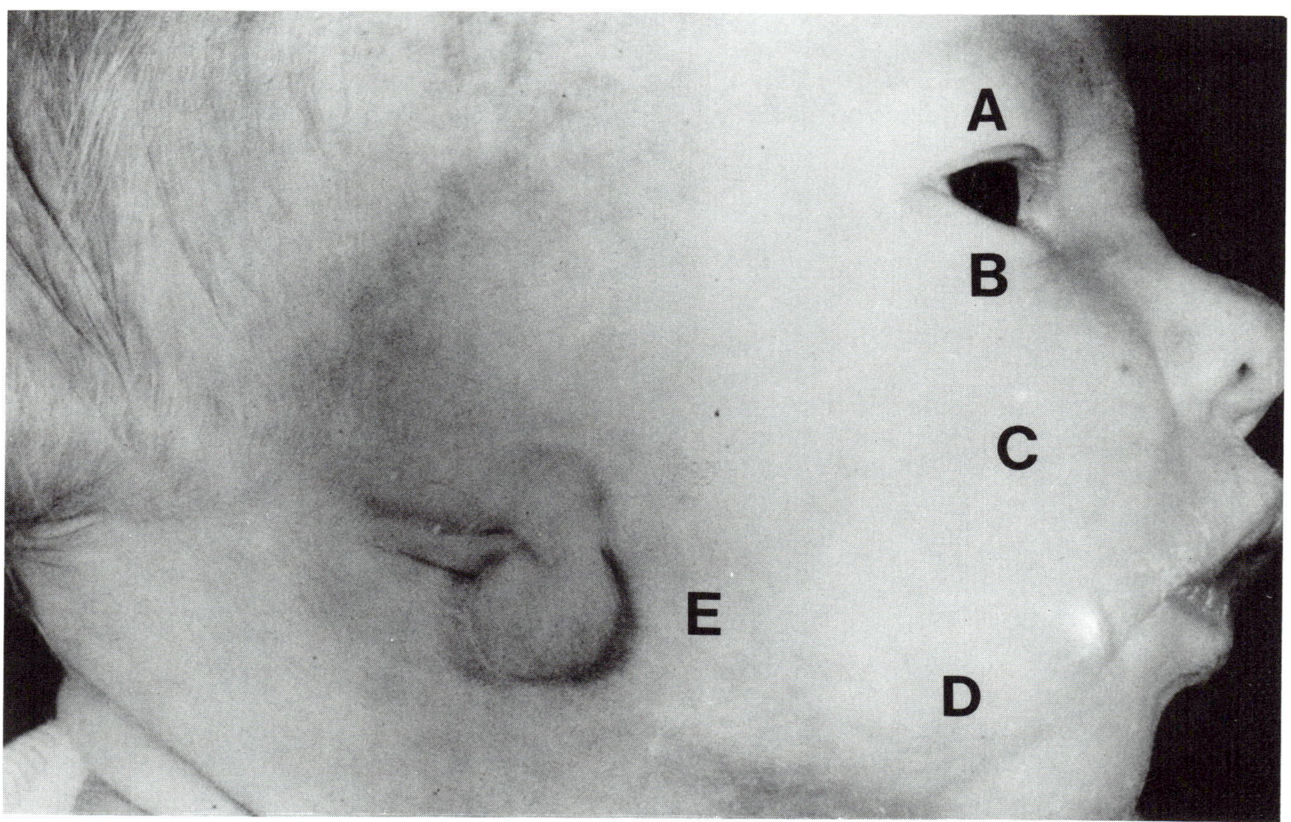

Figure 5.5

222. Hypoplasia of what structures in the embryo would result in this type of defective auricle?
    (A)   first pharyngeal arch ectoderm/mesoderm
    (B)   second pharyngeal arch ectoderm/mesoderm
    (C)   third pharyngeal arch ectoderm/mesoderm
    (D)   fourth pharyngeal arch ectoderm/mesoderm
    (E)   otic vesicle and capsule

223. Upon examination you discover that the external auditory meatus also is absent. This structure normally develops from

    (A)   first pharyngeal groove ectoderm
    (B)   first pharyngeal pouch endoderm
    (C)   second pharyngeal groove ectoderm
    (D)   second pharyngeal pouch endoderm
    (E)   first and second pharyngeal arch mesoderm

Using the letters in **Figure 5.5** and the following key, answer Items 224-227.

    (A)   Maxillary part of the first pharyngeal arch
    (B)   Mandibular part of the first pharyngeal arch
    (C)   Second pharyngeal arch
    (D)   Third pharyngeal arch
    (E)   None of the above

224. The muscle located deep to the skin at A and B and its motor nerve are associated with what embryonic structure?

225. Sensations from the skin at C are conducted by the nerve in what embryonic structure?

226. The bone deep to D forms from the mesenchyme in what embryonic structure?

227. The salivary gland deep to E develops from what embryonic structure?

ANSWERS AND TUTORIAL ON ITEMS 222-227

The answers are: 222-B; 223-A; 224-C; 225-A; 226-B; 227-E. Since the posterior part of the auricle normally forms from three ectodermal hillocks on the lateral surface of the second pharyngeal arch and the underlying mesoderm, the correct answer to Item 222 is B. The anterior part of the auricle similarly forms from three **auricular hillocks** on the lateral surface of the first pharyngeal arch. The ectodermally lined groove between the first and second pharyngeal arches gives rise to the **external auditory canal**.

    The **orbicularis oculi** muscle is located deep to A and B and is innervated by the **facial nerve**. The muscle and nerve are associated with the second pharyngeal arch. The skin at C is derived from the surface ectoderm on the maxillary part of the first pharyngeal arch and is innervated by the infraorbital branches of the **maxillary nerve**. The mandible is located deep to D and develops from tissue in the mandibular part of the first pharyngeal arch. The **parotid salivary gland** that is located deep to E arises in the sixth week from the oral ectodermal lining near the angle of the **stomodeum** (primordial mouth) that is the junction area of the maxillary and mandibular parts of the first arch. Therefore, the best answer to Item 227 is E.

Items 228-235

Indicate the embryonic derivative of each of the definitive structures in Items 228-235 using the following key. Answers may be used once, more than once, or not at all.

        (A)   Derivative of the third pharyngeal groove, pouch, or arch
        (B)   Derivative of the fourth pharyngeal groove, pouch, or arch
        (C)   Both
        (D)   Neither

228.   parathyroid gland

229.   greater horn of the hyoid bone

230.   constrictor muscles of the pharynx

231.   thyroid gland

232.   thymus

233.   muscles of facial expression

234.   muscles of the larynx

235.   stapedius muscle

ANSWERS AND TUTORIAL ON ITEMS 228-235

The answers are: **228-C; 229-A; 230-B; 231-D; 232-A; 233-D; 234-B; 235-D**. The **superior parathyroid gland** develops from the **fourth pharyngeal pouch** and the **inferior** one develops from the **third pouch**. The **greater horn of the hyoid bone** forms from the **third arch cartilage**. The **pharyngeal constrictor** muscles develop from **fourth arch mesenchyme** and are innervated by the nerve of that arch, CN X. Although the **thyroid gland** has a lobe on each side of the midline it develops from a midline diverticulum of pharyngeal (foregut) endoderm. The **thymus** is a derivative of the **third pharyngeal pouch** (A). The muscles of **facial expression** (e.g., orbicularis oris, buccinator, platysma) are innervated by the facial nerve. They are derivatives of **second arch mesenchyme**. The muscles of the **larynx** form from the mesenchyme of the **fourth pharyngeal arch**. The **stapedius muscle** forms from **second pharyngeal arch mesenchyme**.

Items 236-247

Examine the labeled scanning electron micrograph of a human embryo in **Figure 5.6** below and then match each of the statements below with the **MOST** appropriate lettered structure in the micrograph. Lettered structures may be used once, more than once, or not at all.

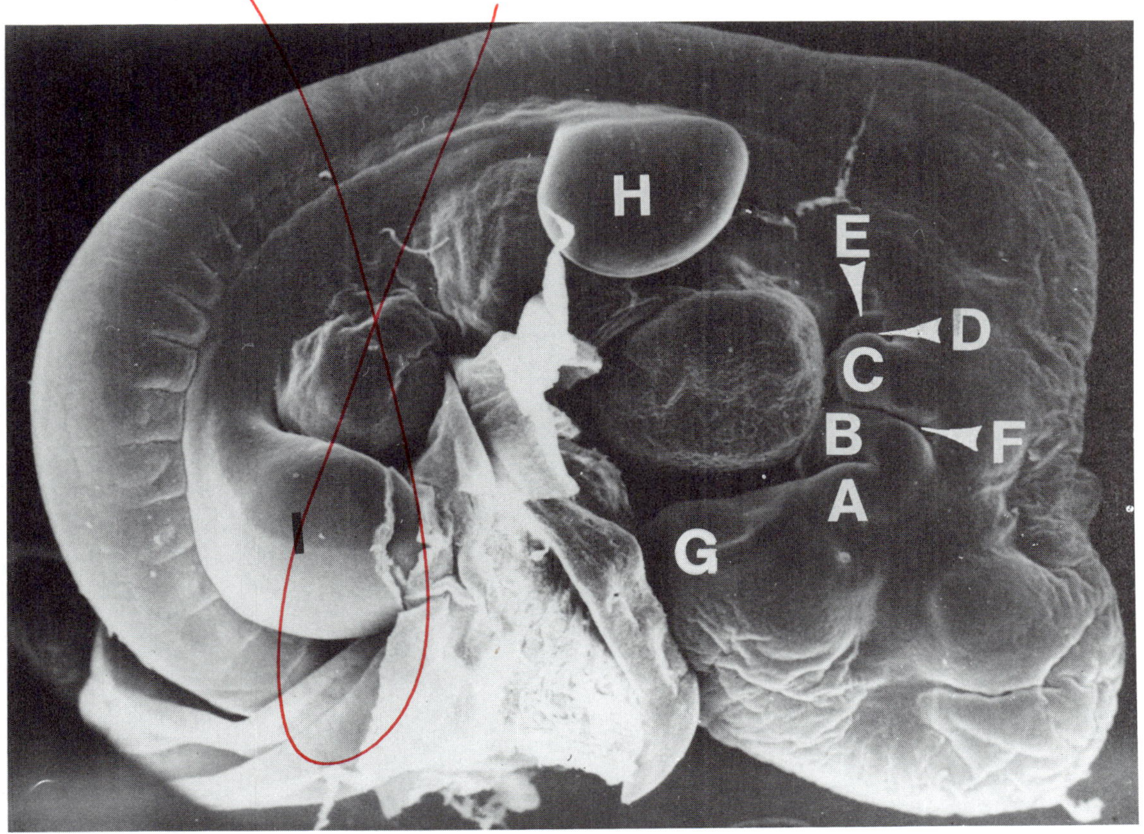

Figure 5.6

236.   Developmental anomalies in this structure cause the mandibular hypoplasia seen in mandibulofacial dysostosis (Treacher Collins syndrome).

237.   The mesenchyme in this structure forms the maxilla.

238.   The aortic arch in this pharyngeal arch forms the hyoid and stapedial arteries in the adult.

239.   This structure forms the external auditory canal in the adult.

240.   Meckel's cartilage forms in the mesenchyme of this structure.

241.   The aortic arch in this structure gives rise to part of the internal carotid artery.

242.   The masseter and temporalis muscles differentiate in the mesoderm of this structure.

243.   Micrognathia results from hypoplasia of this structure.

244.   The styloid process, stylohyoid ligament and lesser horn of the hyoid bone are remnants of the cartilage in this structure.

245.   The central portion of the forehead and the frontal bone is formed in this structure.

Choose the **BEST** response.

246.   All of the following statements about pharyngeal arches are true **EXCEPT**:

    (A)   The arches contain mesenchyme derived from neural crest cells.
    (B)   Skeletal structures in the arches are derived from neural crest cells.
    (C)   Arch-derived skeletal muscles arise *in situ* from somitomeres.
    (D)   The amount of adult tissue derived from each arch is equal.
    (E)   The vascular endothelium is derived from the resident mesoderm.

247.   The laryngeal cartilages form in which pharyngeal structure?

    (A)   second pouch
    (B)   second arch mesenchyme
    (C)   third pouch
    (D)   third arch mesenchyme
    (E)   fourth arch mesenchyme

ANSWERS AND TUTORIAL ON ITEMS 236-247

The answers are: **236-B; 237-A; 238-C; 239-F; 240-B; 241-D; 242-B; 243-B; 244-C; 245-G; 246-D; 247-E. Figure 5.6** is a scanning electron micrograph of a human embryo of approximately 5 weeks after fertilization. The first pharyngeal arch is divided into a **maxillary prominence** (A) and a **mandibular prominence** (B). The **first pharyngeal groove** (F) is clearly visible between the first and **second** (C) **pharyngeal arches**.

The mesenchyme in the maxillary prominence will form the maxilla. The mesenchyme in the mandibular prominence will form Meckel's cartilage and eventually the mandible. Micrognathia would result from abnormal development of the mandible. **Mandibulofacial dysostosis** (first arch syndrome, Treacher Collins syndrome) is caused by an autosomal dominant mutation with a sex ratio of 1:1. This mutation causes developmental anomalies in derivatives of the first pharyngeal arch. The major clinical findings associated with this anomaly are downward slanting palpebral fissures, malar hypoplasia, microtia and conductive hearing loss due to abnormalities in the auditory ossicles and external auditory meatus. This hearing loss is often misdiagnosed as mental retardation which is not a characteristic of people with mandibulofacial dysostosis. The muscles of mastication, e.g., masseter and temporalis muscles, are formed in first arch mesenchyme and are innervated by the mandibular division of the Vth (trigeminal) cranial nerve. The aortic arches of the first and second pharyngeal arches undergo extensive degenerative changes during development but persist in the adult as the maxillary artery (first arch) and hyoid and stapedial artery (second arch). The first pharyngeal groove gives rise to the external auditory canal while the **first pharyngeal pouch** forms the auditory tube.

The **second pharyngeal arch** (C) forms the muscles of facial expression, e.g., the orbicularis oris and orbicularis oculi. The muscles of facial expression are innervated by the VIIth cranial nerve (facial nerve). The **second aortic arch** forms the hyoid and stapedial arteries. The second (D) and more caudal pharyngeal grooves disappear due to caudal overgrowth by the second pharyngeal arch. The **second pharyngeal pouch** forms the fossa of the palatine tonsils.

The **third pharyngeal arch** (E) is also visible. Both the third and fourth pouches contribute to the parathyroid glands while the thymus is derived from the third pharyngeal arch only. The **third aortic arch** forms a portion of the internal carotid artery. The rest of the internal carotid artery is derived from the dorsal aorta. The fourth pharyngeal arch has disappeared in this embryo. The mesoderm of the fourth pharyngeal arch forms the laryngeal cartilages.

Tissues traditionally thought of as solely mesodermal in origin are also derived from neural crest cells in the head and neck. Head mesenchyme and pharyngeal arch cartilages are derived from neural crest. Skeletal muscles and vascular endothelium apparently arise from somite-like (not completely segmented) masses of paraxial mesoderm called **somitomeres**.

The **frontal prominence** (G) will become the forehead. The frontal bone forms in the mesenchyme of the frontal prominence by intramembranous ossification, i.e., without a preformed cartilaginous model. The forelimb bud (H) and hindlimb bud (I) are both visible.

For the convenience of the reader, the basic adult derivatives of the different components of the pharyngeal apparatus are summarized in **Table 5.1** below:

TABLE 5.1

SUMMARY OF ADULT STRUCTURES DERIVED FROM PHARYNGEAL ARCHES

Pharyngeal Arch	Arch Derivatives			Pouch Derivatives	Groove Derivatives	Nerve Supply
	Muscles	Skeletal Structures	Ligaments			
First (mandibular)	Mastication muscles Mylohyoid Anterior belly of digastric Tensor tympani Tensor veli palatini	(Meckel's cartilage) Malleus Incus ventral end of mandible	Anterior ligament of malleus Sphenomandibular ligament	Tubotympanic recess (tympanic membrane, tympanic cavity, mastoid antrum, auditory tube)	External auditory canal	V (trigeminal)
Second (hyoid)	Facial expression muscles Stapedius Stylohyoid Posterior belly of digastric	(Reichert's cartilage) Stapes Styloid process Hyoid bone (lesser horn and upper body)	Stylohyoid ligament	Tonsilar fossa	None	VII (facial)
Third	Stylopharyngeus	Hyoid bone (greater horn and lower body)	None	Inferior parathyroid Thymus	None	IX (glossopharyngeal)
Fourth and sixth combined	Cricothyroid Levator veli palatini Constrictors of pharynx Intrinsic muscles of larynx	Laryngeal cartilages (cricoid, thyroid, arytenoid, corniculate, cuneiform)	None	Superior parathyroids Ultimobranchial body (C-cells of thyroid)	None	X (superior and recurrent laryngeal branches of vagus)

You examine the neck of the child shown in **Figure 5.7** and find a small opening in the skin along the anterior border of the sternocleidomastoid muscle (arrow). A small amount of mucoid fluid leaks from the opening. A lateral cervical (branchial) fistula is suspected.

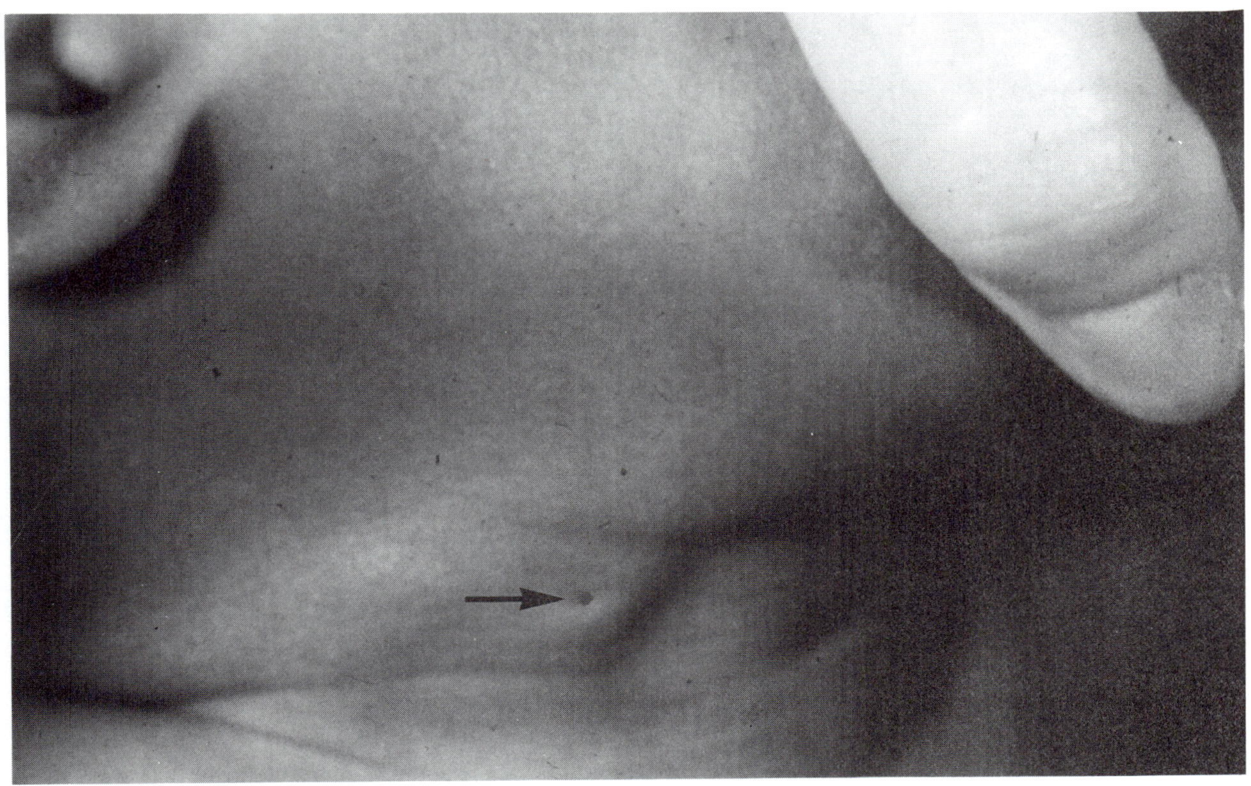

Figure 5.7

Choose the **BEST** response.

248.   Lateral cervical (branchial) fistulas

(A)   usually open externally in the posterior triangle of the neck
(B)   usually open internally in the tonsillar fossa
(C)   usually result from persistence of part of the third pharyngeal groove and pouch
(D)   pass between the common carotid artery and internal jugular vein
(E)   often end blindly near the larynx

249. Lateral cervical (branchial) cysts

(A)   form from the cervical sinus
(B)   are usually located just deep to the body of the mandible
(C)   may form anywhere in the posterior triangle of the neck
(D)   if left alone will usually disappear

ANSWERS AND TUTORIAL ON ITEMS 248 AND 249

The answers are: **248-B; 249-A**. **Lateral cervical (branchial) fistulas** have an external opening in the lateral neck region somewhere along the anterior border of the sternocleidomastoid muscle. Internally, they open in the tonsillar fossa and course between the internal and external carotid arteries. Their removal requires challenging surgery because of their course through the lateral part of the neck. They form when the closing membrane of the second pharyngeal arch degenerates resulting in a communication between the second pharyngeal groove and pouch. **Lateral cervical cysts** form from the cervical sinus. They are often located just inferior to the angle of the mandible but may form anywhere along the anterior border of the **sternocleidomastoid muscle**. The cysts form as a result of fluid and cellular debris accumulating slowly from the original epithelial lining of the groove and pouch. Because of this it is necessary for the epithelial tissue to be removed completely to eliminate the cysts.

Items 250 and 251

Choose the **BEST** response.

250.   The thyroid gland primordium

(A)   appears shortly after the appearance of the pituitary gland primordium
(B)   begins as a down growth of endoderm from the second pharyngeal (branchial) pouch
(C)   may form ectopic thyroid tissue in the tongue
(D)   does not form thyroid follicles until near term
(E)   is nonfunctional in the first trimester

88

251.   All of the following statements about the thyroglossal duct are true **EXCEPT**:

(A)   a portion of it persists as a small, blind pit in the tongue
(B)   cysts may form any where along its course
(C)   when cysts do form they are located usually deep to the trachea
(D)   most thyroglossal duct cysts can be seen by five years of age
(E)   most thyroglossal duct cysts are movable and painless

ANSWERS AND TUTORIAL ON ITEMS 250 AND 251

The answers are: **250-C; 251-C**. The **thyroid gland** is the first endocrine gland to appear during embryonic life. It begins as an endodermal thickening in the floor of the primordial (primitive) pharynx and soon forms a down growth called the **thyroid diverticulum**. At 11 weeks colloid begins to collect in clusters of cells forming a cavity surrounded by a single layer of cells. At this time iodine is concentrated in the thyroid and thyroid hormone can be demonstrated. **Ectopic thyroid tissue** may form from remnants of the **thyroglossal duct** and may be found anywhere between the tongue and neck superior to the definitive thyroid gland.

Thyroglossal duct cysts may form anywhere along the course of the thyroglossal duct. The cysts may form in the tongue or in the median plane of the anterior neck, usually just inferior to the hyoid bone. The cysts are usually painless, progressively enlarging and are easily moveable.

Items 252-256

Near the end of the fourth week, the tongue begins to appear as surface elevations in the floor of the primordial (primitive) pharynx. Match the embryonic rudiment in the items with the **MOST** appropriate description of its contribution in the adult tongue in the answers below. Answers may be used once, more than once, or not at all.

(A)   Anterior two-thirds of the tongue
(B)   Posterior one-third of the tongue
(C)   Both
(D)   Neither

252.   median tongue bud (tuberculum impar)

253.   distal tongue bud (lateral swellings)

254.   hypobranchial eminence

255.   pharyngeal arch mesenchyme forms vessels and connective tissue

256.   occipital somites form most of the muscles

The answers are: **252-D; 253-A; 254-B; 255-C; 256-C**. The first indication of tongue development is the appearance of the **median tongue bud** (tuberculum impar) in the floor of the primordial (primitive) pharynx just rostral to the **foramen cecum**. This bud is a midline elevation of ectoderm. An oval **distal tongue bud** soon forms on each side of the median one, ventromedial to the first pharyngeal arch. The distal tongue buds enlarge rapidly, overgrow the median bud and then merge with one another to form the anterior two-thirds of the tongue (also called the body or oral part of the tongue). The median tongue bud disappears and makes no contribution to the tongue. The posterior one-third of the tongue (also called the root or pharyngeal part of the tongue) initially appears as two midline elevations of endoderm located caudal to the foramen cecum. The more rostral one is named the **copula** and behind it is the **hypobranchial eminence**. The copula is overgrown by the hypobranchial eminence and disappears. The posterior one-third of the tongue forms from the rostral part of the hypobranchial eminence located medial to the third pharyngeal arch. The tongue is innervated by branches of the nerves in the adjacent pharyngeal arches viz., mandibular, facial and glossopharyngeal. Pharyngeal arch mesenchyme forms the blood vessels, lymphatics and connective tissue in the entire tongue. The myoblasts that give rise to most of the tongue muscles are derived from the occipital somites and are innervated by the hypoglossal nerve.

Matching: Match each of the eye abnormalities in the items below with the embryonic structure that is **MOST** likely defective by using the key below. Answers may be used once, more than once, or not at all.

    (A)   Optic vesicle
    (B)   Lens vesicle
    (C)   Lumen of optic vesicle
    (D)   Pupillary membrane
    (E)   None of the above

257.   congenital detachment of the retina

258.   congenital aphakia

259.   congenital glaucoma

260.   anophthalmia

261.   congenital atresia of the pupil

Choose the **BEST** response.

262.   Several spaces are associated with eye development. The space that forms the intraretinal space is

    (A)   optic (choroid) fissure
    (B)   lumen of optic vesicle
    (C)   lumen of optic cup
    (D)   lumen of lens vesicle
    (E)   none of the above

ANSWERS AND TUTORIAL ON ITEMS 257-262

The answers are: **257-C; 258-B; 259-E; 260-A; 261-E; 262-B**. The lumen of the optic cup separates the pigmented (outer) layer of the retina from the neural (inner) layer. These two layers are apposed apically. Recalling that epithelial apices are typically nonadhesive, it is these two layers that separate in **retinal detachment**. Congenital **aphakia** or absence of the lens is the result of an abnormality in or agenesis of the lens vesicle that gives rise to the lens. **Glaucoma** is the abnormal elevation of intraocular pressure leading to blindness. The congenital type is usually the result of abnormal formation of the drainage mechanism of aqueous humor during the fetal period.

Primary **anophthalmia** or absence of the eye results from absence of optic vesicle formation or its arrested development. During the embryonic period the developing lens is invested in a layer of vascular mesenchyme, the anterior portion of which is called the **pupillary membrane**. During the fetal period, the **hyaloid artery** that supplies it usually disappears causing the pupillary membrane to disappear as well. Sometimes the membrane persists as strands of connective tissue over the front of the pupil but rarely interferes with vision. When the entire pupillary membrane persists the condition is called congenital **atresia of the pupil**.

Items 263-266

During a fundoscopic examination you notice that there is a defect in the inferior part of the iris in the patient's right eye. The defect gives the pupil a keyhole appearance. You correctly name the defect a coloboma of the iris.

263. Typically, coloboma iridis results from failure of development of which of the following?

    (A)   lens placode
    (B)   pupillary membrane to reabsorb
    (C)   optic (choroid) fissure to close completely
    (D)   lumen of optic vesicle
    (E)   pars iridica

264. Coloboma iridis

    (A)   usually extends deeper to involve the sclera
    (B)   is frequently hereditary
    (C)   is often fatal
    (D)   is usually caused by retention of the optic sulcus
    (E)   is usually accompanied by coloboma of the lower eyelid

92

265.   All of the following statements are true about microphthalmos **EXCEPT**:

(A)   is severe when it results from arrested development of the optic vesicle
(B)   may be inherited
(C)   may result from maternal exposure to *Toxoplasma gondii*
(D)   may be accompanied by other congenital abnormalities in the head
(E)   is often found in trisomy 21

266.   All of the following statements about congenital cataract is are true **EXCEPT**:

(A)   can be caused by the rubella virus
(B)   can be caused by the congenital galactosemia
(C)   can be inherited as a dominant trait
(D)   is unpreventable with today's technology
(E)   causes blindness

ANSWERS AND TUTORIAL FOR ITEMS 263-266

The answers are: **263-C; 264-B; 265-E; 266-D**. **Coloboma iridis**, a notched out, inferior portion of the iris, is caused by failure of the **optic fissure** to close completely. It is frequently hereditary and, while the defect may extend into the ciliary body and retina, it does not involve the sclera that forms from the mesenchyme surrounding the optic cup. Coloboma iridis is not fatal and is caused by retention of the optic fissure through which the hyaloid vessels enter the optic cup. **Coloboma of the eyelid** results from a localized disturbance in eyelid development and is unrelated to coloboma iridis. In **microphthalmos** the eye may be small and accompanied by other ocular abnormalities or it may be normal in every aspect except that it is small in size. Microphthalmos ia rarely associated with Trisomy 21 (Down syndrome) but is a common feature of **trisomy 13** (Patau syndrome). Congenital cataracts caused by the **rubella virus** could be prevented if all women of reproductive age were immunized against this virus.

<u>**Items 267-272**</u>

Choose the **BEST** response.

267.   Which of the following structures develops from the otocyst?

        (A)   malleus
        (B)   cochlear duct
        (C)   pharyngotympanic (auditory) tube
        (D)   tympanic cavity
        (E)   scala vestibuli

268.   A patient in whom the ventral portion of the otocyst failed to develop bilaterally would have which of the following deficits?

        (A)   deafness
        (B)   unable to maintain balance
        (C)   absent semicircular canals
        (D)   absent auditory ossicles
        (E)   absent auricle

269.   Endoderm of the first pharyngeal pouch contributes to the formation of all the following structures **EXCEPT:**

        (A)   tympanic membrane
        (B)   endolymphatic duct
        (C)   tympanic cavity
        (D)   pharyngotympanic (auditory) tube
        (E)   tubotympanic recess

270.   A poorly formed stapes in the tympanic cavity may result from inadequate development of

        (A)   first pharyngeal arch mesenchyme
        (B)   first pharyngeal pouch endoderm
        (C)   first pharyngeal groove ectoderm
        (D)   second pharyngeal arch mesenchyme
        (E)   second pharyngeal pouch endoderm

271.   All of the following statements about development of the middle ear are true **EXCEPT**:

    (A)   The auditory tube develops from the first pharyngeal pouch.
    (B)   The distal end of the first pharyngeal pouch becomes the tympanic cavity.
    (C)   The ear ossicles and chorda tympani nerve are covered by endodermally derived epithelium.
    (D)   The tensor tympani muscle is derived from first pharyngeal arch mesenchyme.
    (E)   The stapedius muscle is derived from third pharyngeal arch mesenchyme.

272.   All of the following statements concerning deafness are true **EXCEPT**:

    (A)   may result from congenital fixation of the stapes
    (B)   may be associated with Treacher Collins syndrome
    (C)   may result from rubella infection during the critical period of development of the inner ear
    (D)   most types are caused by environmental factors
    (E)   may be associated with other abnormalities of the head

ANSWERS AND TUTORIAL ON ITEMS 267-272

The answers are: **267-B; 268-A; 269-B; 270-D; 271-E; 272-D**. The **otocyst (otic vesicle)** is an ectodermal invagination that is the primordium of the entire **membranous labyrinth** of which the **cochlear duct** is a part. The cochlear duct develops from the ventral portion of the otocyst. Since it contains the receptors for hearing, its failure to develop would result in deafness.

The **endolymphatic sac** is part of the membranous labyrinth and develops from the dorsal utricular portion of the otocyst. All of the other structures listed in Item 269 are derived, to a variable extent, from tubotympanic endoderm of the first pharyngeal pouch.

The **malleus** (which attaches to the medial surface of the tympanic membrane) and incus are formed by the **first pharyngeal arch mesenchyme**. The **stapes** (which attaches to the oval window on the membranous labyrinth) is derived from **second pharyngeal arch mesenchyme**.

The **auditory tube** develops from the **first pharyngeal pouch**. It extends laterally to become continuous with the tympanic cavity this is also a derivative of the first pouch. The ear ossicles and chorda tympani nerve located in the middle ear cavity have a surface covering of endodermally derived epithelium.

The **tensor tympani muscle** attaches to the malleus ear ossicle. Both of these structures are derived from the **first pharyngeal arch**. The **stapedius muscle** is derived from the **second pharyngeal arch** and is innervated by the facial nerve that is the nerve of the second arch.

Most types of congenital deafness are caused by genetic factors but maternal infection by the **rubella virus** during the first trimester may interfere with inner ear development resulting in congenital deafness.

Items 273 and 274

273.   Many congenital birth defects are caused by chromosomal aneuploidy. Abnormalities in the development of the ocular system are **MOST** severe in which of the following aneuploidy conditions?

   (A)   trisomy 21 (Down syndrome)
   (B)   trisomy 13 (Patau syndrome)
   (C)   trisomy 18 (Edwards syndrome)
   (D)   Klinefelter syndrome
   (E)   Turner syndrome

274.   Congenital anomalies in the auditory system are commonly found in which of the following?

   (A)   trisomy 18 (Edwards syndrome)
   (B)   mandibulofacial dysostosis (Treacher Collins syndrome)
   (C)   intrauterine rubella virus infection
   (D)   Potter's syndrome
   (E)   All of the above.

ANSWERS AND TUTORIAL ON ITEMS 273 AND 274

The answers are: **273-B; 274-E.** While different clinical syndromes may include developmental anomalies in the eyes and ears, ocular development is most profoundly disrupted in Trisomy 13 where microphthalmia is common. In **Trisomy 21**, speckling of the iris and oblique palpebral fissures are common but are relatively minor manifestations of the disease.

Different clinical syndromes may affect the development of the auditory system. In **Trisomy 18**, the auricle is dysmorphic in a characteristic way. In mandibulofacial dysostosis, the development of the first pharyngeal arch is altered. Thus, the auricles, external auditory meatus, and middle ear ossicles can be abnormal leading to conduction hearing loss. Rubella infections often target developing neuroepithelium and commonly lead to ocular and auditory abnormalities. **Potter's syndrome** occurs in individuals with bilateral renal agenesis which is fatal shortly after birth. It is associated with characteristic facial features, oligohydramnios and the auricle is commonly abnormal.

96

CHAPTER VI
DEVELOPMENT OF THE
RESPIRATORY AND
CARDIOVASCULAR SYSTEMS

<u>**Items 275-279**</u>

Examine the labeled photomicrograph in **Figure 6.1** below and then choose the **MOST** appropriate descriptive phrase concerning the developmental history of the lettered embryonic structure. Answers may be used once, more than once, or not at all.

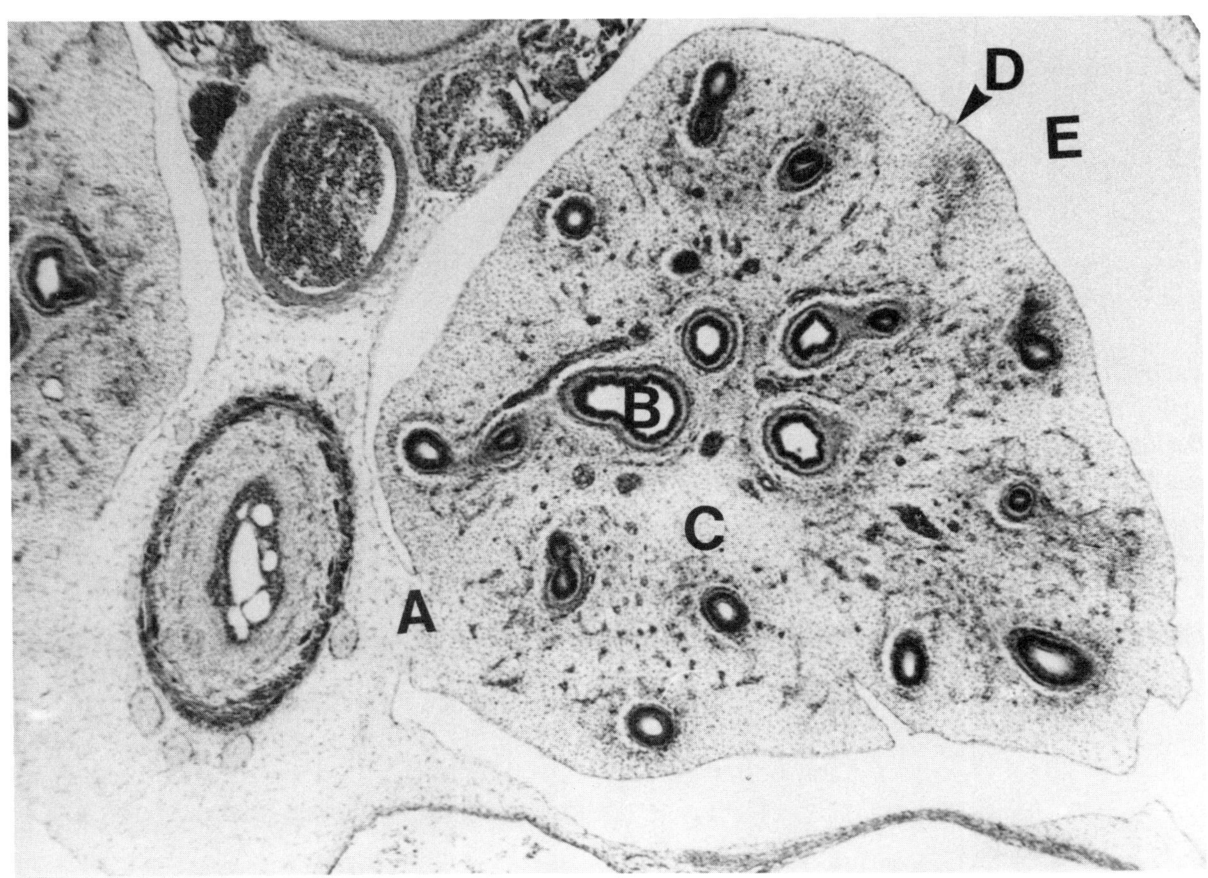

Figure 6.1

275. This structure forms the pulmonary arteries and veins.

276. This structure is in direct communication with the pericardial and peritoneal cavities.

277. This structure will form a lobar bronchus.

278. This structure is a derivative of the respiratory diverticulum.

Choose the **BEST** response.

279. This photomicrograph shows pulmonary development at which stage?

    (A)   pseudoglandular
    (B)   canalicular
    (C)   terminal sac
    (D)   alveolar
    (E)   None of the above.

ANSWERS AND TUTORIAL ON ITEMS 275-279

The answers are: **275-C; 276-E; 277-B; 278-B; 279-A. Figure 6.1** is a photomicrograph of a transverse section through the thorax of a human embryo of approximately 7 weeks gestational age. The lungs form from a reciprocal inductive **epithelial-mesenchymal interaction** between two distinct rudiments: the **respiratory (laryngotracheal) diverticulum** and the **splanchnic mesoderm** (C). The respiratory diverticulum forms as a simple evagination of the foregut which undergoes more than 20 generations of mostly dichotomous branchings. It induces differentiation of mesodermally derived elements e.g., bronchial cartilage, smooth muscle, vasculature, and connective tissue. In turn, the splanchnic mesoderm induces branching of the respiratory diverticulum as well as epithelial differentiation, e.g., pseudostratified, ciliated columnar epithelium in the trachea, bronchi, and bronchioles and a mixture of type I squamous epithelial cells and type II (surfactant-secreting) epithelial cells in alveoli. The respiratory diverticulum forms **ONLY** the epithelial lining of the airways in the trachea, bronchi, bronchioles, and alveoli. Structure B is a primitive lobar bronchus. **ALL** other pulmonary structures are derived from the splanchnic mesoderm. Thus, splanchnic mesoderm forms tracheal and bronchial cartilage, bronchiolar smooth muscle, all the connective tissues of the airways and pulmonary parenchyma, and all the blood and lymphatic vessels in the lungs.

As the respiratory diverticulum grows into the splanchnic mesoderm, the entire developing lung projects into the **intraembryonic coelom** (E) which is the precursor of the pleural cavity as well as the pericardial and peritoneal cavities. At this early stage of development, the intraembryonic coelom is a continuous cavity and is not yet divided into separate pleural, pericardial, and peritoneal cavities. As the heart and lungs grow, and the diaphragm forms, the intraembryonic coelom becomes separated into 4 separate cavities (2 pleural cavities, 1 pericardial cavity, and 1 peritoneal cavity).

During its histological differentiation, the lung passes through 4 distinct phases. The first one is called the **pseudoglandular** (glandular) **period**. It is characterized by extensive branching of the respiratory diverticulum to form the portions of the airway that will ultimately differentiate into the bronchi and bronchioles. This period extends from 5 to 17 weeks of gestation but because of the lack of respiratory bronchioles or alveoli, fetuses born during this phase of pulmonary development are not viable. The second phase is called the **canalicular period**. It is characterized by the growth of the lumen of the airways, the establishment of terminal bronchioles, and development of an extensive vasculature in the splanchnic mesoderm. It extends from 13 to 25 weeks. By the end of the canalicular phase, respiratory bronchioles have formed and enough respiration is possible that premature fetuses born at this time have an increased chance of survival. The third phase is called the **terminal sac (saccular) period**. The epithelium lining the terminal airways becomes very thin and type II (surfactant-secreting) cells differentiate. This period extends from 24 weeks to term. There is extensive postnatal growth and development of the lungs referred to as the **alveolar period**. This period extends from the late fetal period to about eight years. The terminal sacs become alveolar ducts and mature alveoli form as bulges in their walls. The epithelial lining becomes extremely thin. The number and size of the alveoli increase after birth causing the lungs to enlarge.

Items 280-283

A newborn infant is delivered at 30 weeks since the last menstrual period. At birth, the infant weighs 1500 gm but otherwise appears normal. Soon after birth the infant becomes cyanotic and breathes with a grunting noise. Chest X-rays reveal dense lungs with significant atelectasis.

280. The **MOST** likely diagnosis associated with these symptoms is

    (A)   congenital diaphragmatic hernia
    (B)   coarctation of the aorta
    (C)   tetralogy of Fallot
    (D)   renal agenesis
    (E)   respiratory distress syndrome

281. The **MOST** significant predictive prenatal test for this disease would be

    (A)   ultrasonography
    (B)   X-rays
    (C)   measurement of amniotic fluid lecithin/sphingomyelin ratio
    (D)   karyotype analysis
    (E)   measurement of maternal serum α-fetoprotein

282. The cell type **MOST** directly involved in the etiology of this condition is

    (A)   type II pneumocyte
    (B)   macrophage
    (C)   ciliated cell
    (D)   goblet cell
    (E)   erythrocyte

283. In threatened premature delivery, the **MOST** useful drug class for treatment of the mother to accelerate fetal lung maturation is

    (A)   antibiotic
    (B)   antimetabolite
    (C)   neuroleptic
    (D)   steroid
    (E)   diuretic

ANSWERS AND TUTORIAL ON ITEMS 280–283

The answers are: **280-E; 281-C; 282-A; 283-D**. The leading cause of death among premature infants is **respiratory distress syndrome**. A fetus 30 weeks from the last menstrual period is actually 28 weeks old. At 28 weeks **type II pneumocytes** first differentiate and begin to secrete surfactant. The most abundant phospholipid in **surfactant** is **lecithin**. The sphingomyelin levels in amniotic fluid are relatively constant throughout gestation but the lecithin levels begin to rise slowly at 28 weeks. Once many type II cells differentiate, they secrete substantial quantities of surfactant which is transported to the amniotic fluid by fetal respiratory movements. Due to the increasing surfactant secretion, the lecithin/sphingomyelin ratio increases. Type II cell differentiation is a steroid dependent process. Treatment of women with certain steroid drugs will stimulate precocious differentiation of type II cells and thus reduce the risk of respiratory distress syndrome in premature infants.

Items 284-287

You are asked to evaluate a newborn infant who has had a low Apgar score and still has inadequate color on weak crying 2 hours after birth, despite suctioning and Ambulatory Mechanical Breathing Unit (AMBU) ventilation. AP (**Figure 6.2A**) and lateral (**Figure 6.2B**) chest X-rays are obtained and brought to you as you listen for breath sounds in the chest.

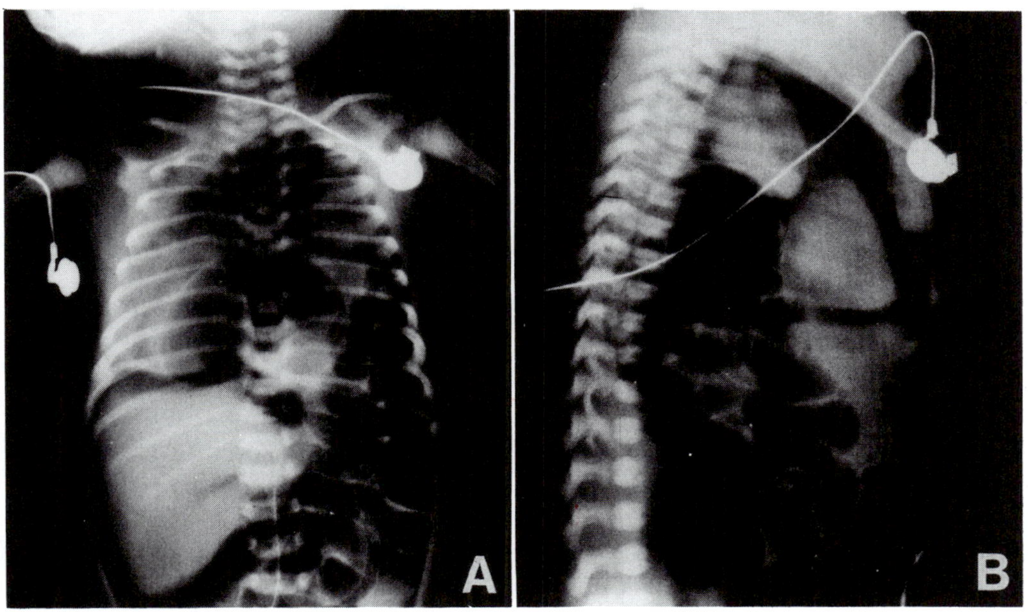

Figure 6.2

284. The **MOST** likely diagnosis is

    (A)   pulmonary sequestration
    (B)   meconium aspiration
    (C)   tracheoesophageal fistula
    (D)   diaphragmatic hernia
    (E)   tension pneumothorax

285. The next **MOST** likely complication to develop if this problem is untreated is

    (A)   hemorrhage
    (B)   perforation
    (C)   hypoxic arrest
    (D)   short gut syndrome
    (E)   organizing pneumonia

286. This congenital anomaly is due to failure of proper morphogenesis of which of the following structures?

    (A)   pericardioperitoneal fold
    (B)   pleuroperitoneal fold
    (C)   pleuropericardial fold
    (D)   dorsal mesoesophagus
    (E)   septum transversum

287. Survival of infants born with a diaphragmatic hernia is **MOST** dependent on

    (A)   time to recognition of disease
    (B)   degree of pulmonary hypoplasia
    (C)   adequacy of antibiotic coverage
    (D)   capacitance of abdominal cavity
    (E)   oxygen toxicity

ANSWERS AND TUTORIAL ON ITEMS 284-287

The answers are: **284-D; 285-C; 286-B; 287-B**. The chest X-ray films in **Figure 6.2** show a newborn with **congenital diaphragmatic hernia** (of Bochdalek). Normally, the diaphragm forms by the fusion of 4 rudiments:

1) septum transversum
2) dorsal mesoesophagus
3) **pleuroperitoneal folds**
4) muscular ingrowth from the body wall

When the pleuroperitoneal folds fail to form properly, more often on the left than right side, a large opening called the **foramen of Bochdalek** develops in the posterolateral region of the diaphragm.

    In this patient, substantial small bowel loops are present in the thoracic cavity through the incomplete separation of the pleural and peritoneal cavities by the congenital diaphragmatic defect. The presence of the bowel in the chest hinders formation of the lungs leading to **pulmonary hypoplasia** which decreases the alveolar surface area for gas exchange. In some instances the developing heart is shifted to the right and may also exhibit congenital defects. Even with suctioning, positive pressure ventilation, and intubation, the newborn's skin does not become pink. Because the persistent hypoxia, hypoxic arrest is a likely outcome. Since the defect is large, incarceration of abdominal viscera within the chest is unlikely. Although the viscera that have become resident in the chest have "lost domain" and the peritoneal capacitance may be decreased, the rate-limiting feature of diaphragmatic hernia is the concomitant failure of pulmonary development. The **degree of pulmonary hypoplasia** is often the feature that determines survival in such infants in whom abdominal viscera are returned to the abdomen and the diaphragmatic hernia is repaired surgically. The adequacy of the lungs to maintain oxygenated blood in the newborn is mainly a function of the degree of pulmonary hypoplasia. Sometimes this condition may require extrapulmonary oxygenation such as with extracorporeal membrane oxygenator use.

Items 288-294

Match the adult structures in the items below with the **MOST** appropriate embryonic precursor forming the adult structure. Answers may be used once, more than once, or not at all.

(A)   Respiratory diverticulum
(B)   Splanchnic mesoderm
(C)   Both
(D)   Neither

288.   bronchial mucosa

289.   tracheal goblet cell

290.   bronchiolar smooth muscle

291.   type I pneumocytes

292.   alveolar macrophages

293.   visceral pleura

294.   Clara cells

ANSWERS AND TUTORIAL ON ITEMS 288-284

The answers are: **288-C; 289-A; 290-B; 291-A; 292-D; 293-B; 294-A**. The **respiratory diverticulum** forms from foregut endoderm and gives rise to the entire epithelial lining of the airways and the alveoli. Thus, it forms goblet cells in the trachea, Clara cells in the bronchioles, and type I and type II pneumocytes in the alveoli. The mucosa of the bronchi consists of an epithelium derived from the respiratory diverticulum and a subjacent connective tissue domain derived from splanchnic mesoderm.

The **splanchnic mesoderm** forms all of the mural connective tissue, smooth muscle, and vascular elements in the airways and the connective tissue and capillaries in the interalveolar septa. It also forms the connective tissue capsule of the lungs and the visceral pleura including the visceral pleural mesothelium. Alveolar macrophages are part of the mononuclear phagocyte system and as such are derived from **bone marrow**.

Items 295-301

Choose the **BEST** response.

295. The cardiovascular system first appears during which week of embryonic development?

    (A)   first
    (B)   second
    (C)   third
    (D)   fourth
    (E)   fifth

296. The heart is derived from

    (A)   somatic mesoderm
    (B)   septum transversum
    (C)   somites
    (D)   splanchnic mesoderm
    (E)   paraxial mesoderm

297. The endocardial heart tubes develop

    (A)   cranial and lateral to the neural plate
    (B)   in the endoderm
    (C)   dorsal to the oropharyngeal membrane
    (D)   from cervical somites
    (E)   from cranial somitomeres

298. All of the following statements about cardiac jelly are correct **EXCEPT**:

   (A)   It separates the endocardium from the myocardium.
   (B)   It is involved in cellular interactions between the endocardium and the
         myocardium.
   (C)   It is extracellular matrix material.
   (D)   It becomes the subendocardial connective tissue layer of the heart.
   (E)   It gives rise to the epicardium.

299. Neural crest cells give rise to which component of the developing heart?

   (A)   cardiac muscle
   (B)   endothelium
   (C)   pericardium
   (D)   autonomic ganglia and endocardial cushions
   (E)   None of the above.

300. The myocardial layer of the heart tube develops from

   (A)   endocardial tubes
   (B)   dorsal mesocardium
   (C)   cardiac jelly
   (D)   splanchnic mesoderm
   (E)   septum transversum

301. The dorsal mesocardium

   (A)   disappears partially and helps form the transverse pericardial sinus
   (B)   persists as the fibrous pericardium
   (C)   fuses with the dorsal mesogastrium
   (D)   disappears and helps form the oblique pericardial sinus
   (E)   persists as the endocardium

The answers are: **295-C; 296-D; 297-A; 298-E; 299-D; 300-D; 301-A**. The cardiovascular system forms early in embryonic development and is the first organ system to become functional. At the beginning of the **third week**, groups of cells (angiogenic cell clusters) appear in the wall of the yolk sac, in the chorion and in the connecting stalk mesoderm. Two heart tubes form within a few days and the single fused heart tube begins to beat at the end of the 3rd week. The heart tubes form from **splanchnic mesoderm** within the **cardiogenic region** cranial to the oropharyngeal membrane and beneath the intraembryonic coelom.

On day 19 of development a pair of vascular elements called the endocardial tubes develop in the cardiogenic region of the embryo. This region is a horseshoe-shaped zone of splanchnopleuric mesoderm located cranial and lateral to the neural plate on the embryonic disc. The cranial and lateral folding of the embryo brings the two lateral endocardial tubes into the thoracic region where they meet in the midline and fuse to form a single primary heart tube.

Cardiac jelly is a relatively thick layer of extracellular matrix material that lies between endocardium and myocardium during early heart formation. As with extracellular matrix in many places, it is probably involved in cellular interactions during heart development.

Experimental studies on lower animals show that **neural crest** cells form the cardiac ganglia, the smooth muscle in the great vessels, the endocardial cushions, and play a major role in bulbotruncal (conotruncal) septation.

The **myocardium** or heart muscle arises from **splanchnic mesoderm** near the heart tubes in the cardiogenic region. The endocardium is also derived from cardiogenic splanchnic mesoderm, where angiogenic cell clusters form the heart tubes. The epicardium develops from the splanchnic mesoderm lining the intraembryonic coelom.

Early in heart development, following head folding, the heart tube is suspended within the pericardial cavity by the dorsal mesocardium. When heart folding occurs the dorsal mesocardium ruptures. The ruptured area becomes the **transverse pericardial sinus** located behind the great arteries. The transverse sinus separates the arterial and venous "ends" of the heart just as the ruptured myocardium does in the embryo.

Match each of the descriptions relating to heart development in the items below with the **MOST** appropriate lettered structure in **Figure 6.3** below. Each answer may be used once, more than once, or not at all.

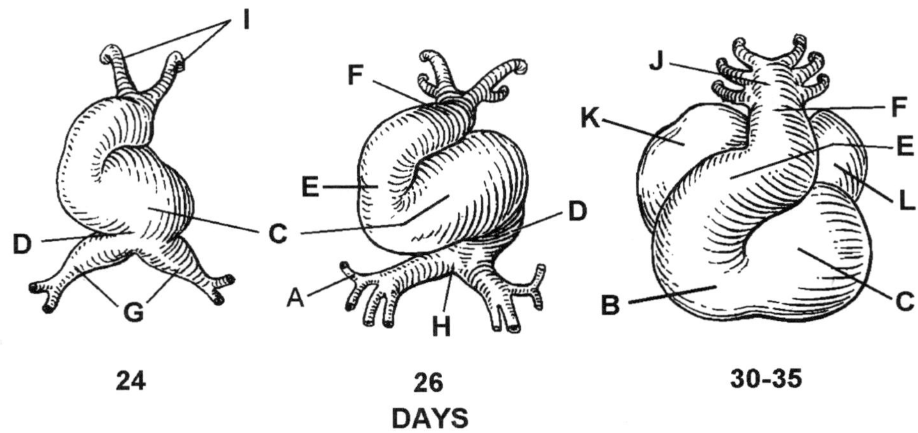

Figure 6.3

302. Gives rise to the ascending aorta and pulmonary trunk.

303. Gives rise to most of the trabeculated part of the right ventricle.

304. Gives rise to the smooth-walled part of the right atrium.

305. Region known as the conus cordis.

306. Represents first aortic arch.

307. Region known as the primitive atrium.

308. Region which receives the two vitelline veins.

The answers are: **302-F; 303-B; 304-H; 305-E; 306-I; 307-D; 308-H. Figure 6.3** illustrates the early development of the human heart. The **sinus venosus** (H) is a symmetric expansion of the caudal end of the primitive heart tube consisting of a median region and a **left and right sinus horn** (G). The **vitelline**, umbilical, and common cardinal veins open into the sinus venosus. The sinus venosus gives rise to the smooth-walled part of the **definitive right atrium**. The part of the right atrial wall with pectinate muscles (anterior to the crista terminalis) and the right and left auricles are derived from the primitive atrium. The **primitive ventricle** (C) gives rise to most of the **definitive left ventricle** and is separated from the primitive atrium by an atrioventricular sulcus. The bulbus cordis differentiates into several structures and is separated from the ventricle by the bulboventricular sulcus. The most caudal region of the **bulbus cordis** (B) forms the trabeculated part of the **right ventricle**. The rest of the bulbus cordis, called the **conus cordis** (E), contributes to the formation of two definitive outflow tracts of the ventricles called the infundibulum of the right ventricle and the aortic vestibule of the left ventricle. The **truncus arteriosus** (F) forms the **ascending aorta** and the **pulmonary trunk**. The **aortic sac** (J) is cranial to the heart tube and connects directly to the **first aortic arches** (I). The other aortic arches will arise from the aortic sac.

Matching. Match the answers listed below with the **MOST** appropriate statement that follows. Answers may be used once, more than once, or not at all.

(A)	Truncal swellings
(B)	Right sinus horn
(C)	Left sinus horn
(D)	Sinus venosus
(E)	Endocardial cushions
(F)	Left venous (sinoatrial) valve
(G)	Right vitelline vein
(H)	Left vitelline vein
(I)	Subendocardial proliferations

309. Gives rise to the bicuspid and tricuspid valves.

310. Endocardial thickenings that line the truncus arteriosus and give rise to a septum separating the aorta and pulmonary trunk.

311. Is reduced and forms the oblique vein (of Marshall) of the left atrium.

312. Give rise to the aortic and pulmonary semilunar valves.

313. Becomes incorporated into the right atrium forming its smooth portion.

314. Gives origin to the septum secundum during formation of the interatrial septum.

ANSWERS AND TUTORIAL ON ITEMS 309-314

The answers are: **309-I; 310-A; 311-C; 312-I; 313-B; 314-F**. The **truncal swellings** (A) are longitudinal thickenings in the endocardium that line the truncus arteriosus. The ridges fuse into a septum that separates the future aortic and pulmonary trunks. Both the **aortic** and **pulmonary semilunar valves** arise from **proliferations of the subendocardial tissue** around the orifices of each trunk. The septum formed by the fused bulbar ridges in the bulbus cordis separates the bulbus cordis into the conus arteriosus of the adult right ventricle and the aortic vestibule of the left ventricle.

The septum which forms the partition between the atrial walls is formed by the **septum primum** and the **septum secundum**. The septum secundum incorporates the primitive **left venous (sinoatrial) valve**.

The **sinus venosus** receives the venous drainage of the embryo. Early on it has two entrance channels, the right and left sinus horns. The **left horn** of the sinus venosus becomes less important about the fifth week as blood is diverted to the right horn. Its remnants in the adult are the **coronary sinus** and the stem of the **oblique vein** (of Marshall) of the left atrium. The right sinus horn (B) becomes incorporated into the right atrium, forming the smooth portion of this chamber.

Endocardial cushions are dorsal and ventral thickenings in the wall of the primitive atrioventricular canal. **Neural crest cells** populate the endocardial cushions. The cushions fuse with one another and separate the primitive atrioventricular canal into right and left portions. The cushions also participate in the formation of the interatrial and interventricular septa. The cusps of the **tricuspid and bicuspid valves** arise from localized **proliferations of subendocardial tissue** around the right and left atrioventricular canals, respectively.

<u>Items 315-320</u>

Choose the **BEST** response.

315.   In the normal fetal circulation, blood from the placenta bypasses the sinusoidal plexus of the liver by way of the

    (A)   sinus venosus
    (B)   ductus venosus
    (C)   cardinal veins
    (D)   vitelline veins
    (E)   umbilical vein

316.   All of the following ligaments in the adult are derived from fetal blood vessels **EXCEPT**:

    (A)   median umbilical ligament
    (B)   round ligament of the liver
    (C)   ligamentum venosum
    (D)   ligamentum arteriosum
    (E)   medial umbilical ligament

110

317.   During embryonic and fetal development

    (A)   pulmonary arteries carry highly oxygenated blood
    (B)   hepatic segment of the inferior vena cava carries highly oxygenated blood
    (C)   umbilical arteries carry highly oxygenated blood
    (D)   All of the above.
    (E)   None of the above.

318.   All of the following structures are remnants of the fetal circulation **EXCEPT**:

    (A)   fossa ovalis of the heart
    (B)   musculi pectinati of the atria
    (C)   oblique vein of the left atrium
    (D)   ligamentum arteriosum
    (E)   ligamentum teres hepatis

319.   All of the following structures are correctly paired **EXCEPT**:

    (A)   right vitelline vein - inferior vena cava
    (B)   left vitelline vein - liver sinusoids
    (C)   right anterior cardinal vein - part of superior vena cava
    (D)   right umbilical vein - definitive umbilical vein
    (E)   left sinus horn - coronary sinus

320.   Which of the following blood vessels contains the **MOST** oxygenated blood during fetal development?

    (A)   superior vena cava
    (B)   right ventricle
    (C)   pulmonary artery
    (D)   ascending aorta
    (E)   umbilical artery

The answers are: **315-B; 316-A; 317-B; 318-B; 319-D; 320-D**. The **ductus venosus** is a shunt that develops between the left umbilical vein and the hepatic segment of the inferior vena cava. It allows highly oxygenated blood from the placenta to bypass the liver sinusoids.

The **median umbilical ligament**, attaching to the upper part of the bladder, is a remnant of the urachus, which itself was a remnant of the allantois. The allantois was a connection between the urinary bladder and the yolk sac, not a fetal blood vessel.

In the embryo blood returns from the caudal half of the body by way of the posterior cardinal veins. Later, the **inferior vena cava** carries blood from the lower body and receives the ductus venosus, the latter carrying umbilical, highly oxygenated blood. The pulmonary and umbilical arteries carry poorly oxygenated blood.

The fossa ovalis, oblique vein, ligamentum arteriosum and ligamentum teres hepatis are all remnants of the fetal circulation. The **pectinate muscles** are a feature of the auricles of each atria and are derived from the wall of the original primitive atrium.

The **right umbilical vein** totally disappears. The definitive, single umbilical vein is the left one. After birth the umbilical vein closes and becomes the ligamentum teres hepatis (round ligament of the liver).

The hepatic segment of the inferior vena cava carries highly oxygenated blood from the ductus venosus to the right atrium. Upon reaching the right atrium, most of this blood passes through the foramen ovale to the left atrium from where it enters the left ventricle and then into the **ascending aorta**. Poorly oxygenated blood entering the right atrium from the superior vena cava courses into the right ventricle and then the pulmonary trunk.

Items 321-325

Choose the **BEST** response.

321.   Which of the following vessels gives rise to the hepatic portal vein?

      (A)   umbilical vein
      (B)   umbilical artery
      (C)   vitelline vein
      (D)   3rd aortic arch
      (E)   6th aortic arch

322.   The vitelline arteries of the embryo give rise to all of the following vessels **EXCEPT**:

      (A)   superior mesenteric artery
      (B)   inferior mesenteric artery
      (C)   celiac artery
      (D)   renal artery

323. During embryonic development the sinusoids of the liver become continuous with

    (A)   vitelline veins
    (B)   cardinal veins
    (C)   umbilical veins
    (D)   All of the above.
    (E)   None of the above.

324. The left brachiocephalic vein is derived from

    (A)   left anterior cardinal vein
    (B)   left posterior cardinal vein
    (C)   a shunt between the left and right posterior cardinal veins
    (D)   left horn of the sinus venosus
    (E)   None of the above.

325. The arch of the azygos vein is derived from

    (A)   posterior cardinal vein
    (B)   anterior cardinal vein
    (C)   common cardinal vein
    (D)   supracardinal vein
    (E)   subcardinal vein

ANSWERS AND TUTORIAL ON ITEMS 321-325

The answers are: **321-C; 322-D; 323-A; 324-E; 325-A**. The **hepatic portal vein** forms when the anastomotic network around the duodenum eventually coalesces to form one vessel. This anastomotic network is derived from the vitelline veins that drain the embryonic gut.

The **vitelline veins** become continuous with the **hepatic sinusoids** as they pass through the developing liver. When the left sinus horn regresses, blood is shunted from the left vitelline vein into the right. The **right vitelline vein** enlarges and ultimately becomes the **hepatic portion of the inferior vena cava**.

The blood vessels that arise in the yolk sac differentiate to form the arteries and veins of the vitelline system. As the yolk sac regresses, the right and left vitelline arterial plexuses coalesce to form a series of major arteries, including the **celiac trunk, superior mesenteric, and inferior mesenteric arteries**. The renal arteries are derived from lateral branches of the abdominal aorta.

When the left vitelline vein regresses the right vitelline vein drains the entire vitelline system and enlarges. The superior portion becomes the terminal portion of the inferior vena cava. While most of the vitelline system inferior to the liver regresses, some portions remain to become the main tributaries of the portal system. The portion of the right vitelline vein inferior to the liver becomes the portal vein and the superior mesenteric vein. The cranial end of the **left brachiocephalic vein** forms from the **left anterior cardinal**, but the lower part develops as a shunt between the left and right anterior cardinal veins. On the right side, the upper part of the superior vena cava is derived from the right anterior cardinal, and the lower part is from the right common cardinal vein. The azygos system of veins are derived from the supracardinal veins, although the **arch of the azygos** is the remnant of the **right posterior cardinal vein**. The point at which the azygos arch empties into the superior vena cava in the adult marks the junction of the right anterior cardinal above with the right common cardinal below.

<u>Items 326-332</u>

Match the lettered structures in **Figure 6.4** below with the **CORRECT** statement that follows. Answers may be used once, more than once, or not at all.

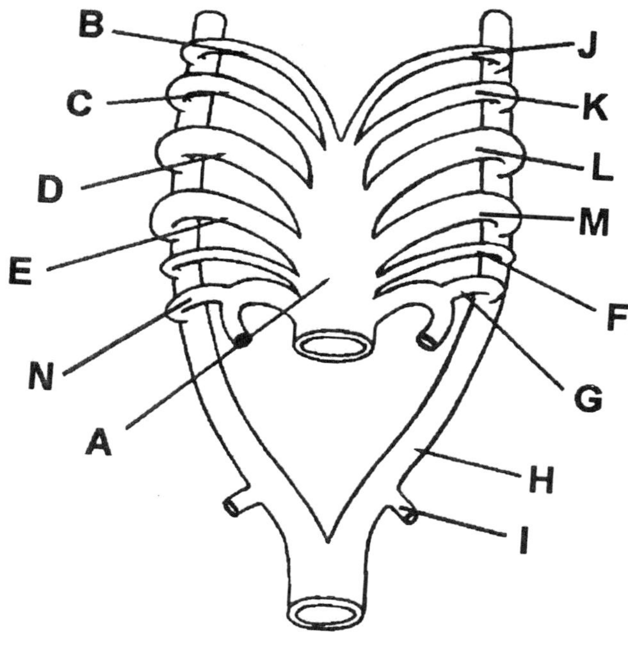

Figure 6.4

326. Gives rise to the ductus arteriosus.

327. Gives rise to right common carotid and right proximal internal carotid artery.

328. Along with the right 7th intersegmental and a portion of the right dorsal aorta, this arch will form the right subclavian artery.

329. In humans, this aortic arch never develops or exists only briefly and does not contribute to any definitive artery.

330. Gives rise to the proximal part of the arch of the aorta.

331. Gives rise to the right maxillary artery.

332. The right recurrent laryngeal nerve loops around the derivative of this aortic arch.

ANSWERS AND TUTORIAL ON ITEMS 326-332

The answers are: **326-G; 327-D; 328-E; 329-F; 330-A; 331-B; 332-E**. During development four well-defined pairs of pharyngeal arches appear externally in humans. Each arch contains an artery called an **aortic arch**. Each arch arises from the **aortic sac** and courses dorsally to the left or right dorsal aorta. The four pairs of arches, numbered 1 through 4, persist to give rise to definitive structures. Based primarily on the ontogeny of lower forms, fifth and sixth pharyngeal arches have been described but in the human embryo they are vestigial

The **truncus arteriosus** and **aortic sac** (A) develop into the **ascending aorta**. The aortic sac also forms the proximal portion of the **arch of the aorta**. The right and left dorsal aortae fuse during the 4th week to form a midline **dorsal aorta**, which later becomes the **descending aorta**. Remnants of aortic arches 1 and 2 give rise to minor vascular elements in the head. The **1st arch** (B, J) forms the terminal segment of the **maxillary artery**. The **2nd arch** (C, K) forms the stapedial artery. **Arches 3 and 4** give rise to important vessels of the head, neck and upper thorax. The **3rd arch** (D, L) will supply blood to the head. Earlier, the head was supplied by the cranial portions of the paired dorsal aortae, but the segments of the dorsal aortae between the 3rd and 4th arches degenerate. The 3rd arch becomes the **common carotid artery** and the **beginning portion of the internal carotid artery**. The distal portion of the internal carotid artery develops from the original dorsal aorta. The **external carotid artery** arises *de novo* from the internal carotid artery.

The **4th arch** (E, M) becomes part of the **arch of the aorta** on the left side and the **subclavian artery** on the right side. Initially, the upper limb receives the seventh cervical intersegmental artery (I), a branch of the dorsal aorta. On the left side, the **left 4th aortic arch** becomes the middle portion of the arch of the aorta. The **left 7th intersegmental artery** becomes the **left subclavian artery**. On the right side, the **right 4th arch** becomes the root of the **right subclavian artery**. The portion of the right dorsal aorta between the right 7th intersegmental artery and the descending aorta degenerates. The **brachiocephalic artery** develops from a portion of the aortic sac.

The precursor of the distal part of the **pulmonary trunk** and the **ductus arteriosus** is often described as a **left sixth arch** (G). The right and left pulmonary arteries originate as branches from the 4th arch but later shift their origin to the pulmonary trunk.

The asymmetrical development of the 4th and 6th arches explains the differing courses of the right and left recurrent laryngeal nerves. On the right side the 5th and 6th arches never form, so the **right recurrent laryngeal nerve** comes to loop around the right subclavian artery, the derivative of the **right 4th arch**. On the left, the 6th arch becomes the ductus arteriosus and thus the **left recurrent laryngeal** loops around this **6th arch** derivative.

Items 333-340

Choose the **BEST** response.

333.   During heart formation, the foramen ovale is an opening in

      (A)   septum primum
      (B)   septum spurium
      (C)   septum secundum
      (D)   truncus arteriosus
      (E)   interventricular septum

334.   The valve of the foramen ovale is formed by

      (A)   septum primum
      (B)   septum secundum
      (C)   truncoconal septa
      (D)   valve of the inferior vena cava
      (E)   septum spurium

335.   Which of the following best describes the period of complete partitioning of the primitive heart into four separate chambers?

      (A)   first trimester
      (B)   second month
      (C)   fifth month
      (D)   first month
      (E)   second trimester

336. The membranous portion of the interventricular septum is formed by

    (A)   bulbar ridges and endocardial cushions
    (B)   truncal ridges
    (C)   septum spurium
    (D)   septum primum
    (E)   septum secundum

337. Partitioning of the atrium is accomplished by formation of

    (A)   septum primum
    (B)   septum secundum
    (C)   Both A and B above.
    (D)   Neither A nor B above.

338. The foramen secundum of the embryonic heart is an opening in

    (A)   septum primum
    (B)   septum secundum
    (C)   interventricular septum
    (D)   endocardial cushions
    (E)   septum spurium

339. The function of the interatrial septum is to

    (A)   completely separate the two atria after birth
    (B)   completely separate the two atria during fetal development
    (C)   act as a valve allowing unidirectional shunting of blood during fetal
           development
    (D)   All of the above.
    (E)   Only A and C above.

340. Closure of the valve of the foramen ovale is accomplished by

    (A)   increased pressure in the right atrium relative to left
    (B)   increased vascular resistance in the lungs
    (C)   closure of the ductus arteriosus
    (D)   increased oxygen content of the blood
    (E)   increase in pressure in the left atrium relative to the right

The answers are: **333-C; 334-A; 335-B; 336-A; 337-C; 338-A; 339-E; 340-E**. The partitioning of the heart begins to occur during the 4th week of development and continues until the end of the embryonic period. **Two septa** make their appearance during this time. They divide the common atrium into left and right atria. The septa are named according to the sequence of their appearance with the first septum named the **septum primum** that precedes slightly the formation of the second septum, the **septum secundum**. The function of the conjoint interatrial septum is to allow continuous shunting of blood from the right to the left atrium during fetal life but close off this shunt after birth.

At the end of the 4th week of development, the **septum primum** grows downward from the dorsocranial roof of the primitive atrium. The septum grows toward the endocardial cushions. The opening between the septum primum and the atrioventricular canal is the **foramen (ostium) primum**. Eventually the septum primum will fuse with the endocardial cushions and obliterate the foramen primum. Failure of closure of the foramen primum is classified as a **foramen (ostium) primum type of atrial septal defect**. Just before the foramen primum disappears, a second foramen, the **foramen (ostium) secundum** forms in the superior region of the septum primum. The second opening permits continuous right-to-left shunting of blood in the fetal heart.

During the 5th week, a thick, muscular second septum named the **septum secundum** grows from the dorsocranial roof of the atrium just to the right of the septum primum. The septum secundum normally does not grow far enough to fuse with the endocardial cushions. An oval-shaped opening remains below through the septum secundum named the **foramen ovale**. The upper portion of the septum primum is resorbed but the remaining lower portion becomes the **valve of the foramen ovale**. Normally, closure of the valve of the foramen after birth results from the increase in the pressure within the left atrium relative to the right atrium.

Items 341-346

Choose the **BEST** response.

341.　An ostium (foramen) secundum defect

  (A) is characterized by a large opening between the left and right atria
  (B) may be caused by excessive resorption of the septum primum
  (C) may be caused by inadequate development of the septum secundum
  (D) may be accompanied by intracardiac shunting of blood
  (E) All of the above.

342. Which of the follow cardiac abnormalities would have the least serious clinical
problems?

    (A)   patent ductus arteriosus
    (B)   mitral valve stenosis
    (C)   atrial septal defect
    (D)   dextrocardia or right-sided heart
    (E)   pulmonary trunk stenosis

343. In tricuspid atresia, usually there is

    (A)   a patent foramen ovale
    (B)   a ventricular septal defect
    (C)   an underdeveloped right ventricle
    (D)   hypertrophy of the left ventricle
    (E)   All of the above.

344. In atresia of the pulmonary semilunar valve

    (A)   a patent foramen ovale is the only outlet for blood from the right side of the
heart
    (B)   the left atrium is markedly underdeveloped
    (C)   a patent ductus arteriosus offers a route of blood flow to the lungs
    (D)   All of the above.
    (E)   Only A and C above.

345. In atresia of the aortic semilunar valve, blood passes into the aorta through the

    (A)   ductus venosus
    (B)   umbilical vein
    (C)   ligamentum arteriosum
    (D)   vitelline vein
    (E)   patent ductus arteriosus

346. The bulbotruncal ridges may fail to undergo normal spiralling during development.
This would **MOST** likely result in

    (A)   transposition of the great vessels
    (B)   ventricular septal defect
    (C)   atrial septal defect as the only defect
    (D)   tetralogy of Fallot
    (E)   patent ductus arteriosus as the only defect

ANSWERS AND TUTORIAL ON ITEMS 341-346

The answers are: **341-E; 342-D; 343-E; 344-E; 345-E; 346-A**. An **ostium** (foramen) **secundum defect** is a large opening between the atria that can be caused either by **excessive resorption of the septum primum** or **inadequate development of the septum secundum**. Depending on the size of the defect, there can be **shunting of the blood** between the atria.

In the case of **dextrocardia**, there is transposition so that the heart and great vessels are mirror images of the normal arrangement. This is a benign situation that is often accompanied by transposition of abdominal viscera, as well. As long as there are no other defects, the patient may never notice the abnormality.

Tricuspid atresia is always accompanied by a patent foramen ovale, a ventricular septal defect, an underdeveloped right ventricle, and a hypertrophic left ventricle.

When the **pulmonary valve is atretic**, the only way for blood to get out of the right side of the heart is through a **patent foramen ovale**. Conversely, the only way for blood to get to the lungs is by way of a patent ductus arteriosus. Because the right ventricle is not pumping blood, it becomes markedly hypoplastic. There is no overriding aorta.

For a patient with an **atretic aortic valve** to survive, there must be a **patent ductus arteriosus** in order for blood to reach the systemic circulation.

The **truncal ridges** or swellings are the progenitors of the aorticopulmonary septum within the truncus arteriosus. During development they undergo a spiralling that places the aorta and pulmonary trunks in proper orientation. If they fail to spiral, **transposition of the great vessels** occurs where the aorta arises from the right ventricle and the pulmonary trunk arises from the left ventricle.

Items 347-349

A 47-year-old man presents with complaints of fatigue. On auscultation you hear a distinct heart murmur (abnormal heart sound). Radiological examination reveals enlargement of the right atrium and right ventricle.

Choose the **BEST** response.

347. Given just this information, which of the following congenital heart malformations is **MOST** likely present in this patient?

    (A)   tetralogy of Fallot
    (B)   primum type atrial septal defect
    (C)   ventricular septal defect
    (D)   patent ductus arteriosus
    (E)   secundum type atrial septal defect

120

348. You would expect the patient above to have which type of blood shunting?

    (A)   right-to-left
    (B)   left-to-right
    (C)   alternating between right-to-left, then left-to-right
    (D)   no shunting of blood

349. The defect in this patient could be the result of

    (A)   defective formation of the septum secundum
    (B)   excessive resorption of the septum primum
    (C)   Both A and B above.
    (D)   Neither A nor B above.

ANSWERS AND TUTORIAL ON ITEMS 347-349

The answers are: **347-E; 348-B; 349-C**. A **foramen (ostium) secundum defect** is the **most common type of congenital heart malformation seen in adults**. This is because the patient may be asymptomatic for 30 or 40 years. The patient will only begin to notice increased fatigability and other symptoms when the right atrium and right ventricle begin to respond to the overloading. A foramen secundum defect is a large opening between the atria that can be caused either by **excessive resorption of the septum primum** or **inadequate development of the septum secundum**. Depending on the size of the defect, there can be **shunting** of the blood between the atria in a **left-to-right direction** because of the greater pressure within the left atrium.

Items 350 and 351

A 42-year-old woman complains of recent difficulty in swallowing. During a routine physical exam, you note that the blood pressure of the left arm is 130 over 84 and the wrist pulse is about 72 beats per minute. You find that the blood pressure in the right arm is 71 over 48 and a pulse is barely detectable.

Choose the **BEST** response.

350.   What congenital vascular anomaly could explain these findings?

    (A)   patent ductus arteriosus
    (B)   anomalous right subclavian artery
    (C)   ventricular septal defect
    (D)   dextrocardia
    (E)   atrial septal defect

351.   The anomaly would be caused embryologically by the abnormal regression of the

    (A)   aortic sac
    (B)   1st aortic arch
    (C)   2nd aortic arch
    (D)   right 4th aortic arch
    (E)   left 4th aortic arch

ANSWERS AND TUTORIAL ON ITEMS 350 AND 351

The answers are: **350-B; 351-D**. The right subclavian artery is normally formed by the right fourth aortic arch, the proximal part of the right dorsal aorta and the right 7th intersegmental artery. If the **right 4th arch** undergoes **abnormal regression**, the right 7th intersegmental artery connects to the aorta and forms the right subclavian artery. When the aorta assumes its final position, the abnormal right subclavian artery takes origin distal to the left subclavian artery and passes dorsal to the esophagus. The esophagus is thus compressed resulting in dysphagia (difficulty swallowing). Conversely, the esophagus may compress the aberrant subclavian artery and thus reduce blood flow to the right arm, giving a reduced pulse and blood pressure reading.

Items 352-354

A newborn girl is delivered with clinically evident cyanosis. A chest film shows that there is a reduction in pulmonary blood flow and the heart is small for a child of her size. An EKG indicates there is right ventricular hypertrophy.

Choose the **BEST** response.

352.   Which of the following conditions is this child **MOST** likely to have?

   (A)   secundum type atrial septal defect
   (B)   coarctation of the aorta
   (C)   tetralogy of Fallot
   (D)   dextrocardia
   (E)   ventricular septal defect

353.   In addition to right ventricular hypertrophy, the condition this child has would also include all of the following abnormalities **EXCEPT**:

   (A)   pulmonary stenosis
   (B)   rightward displacement of the aorta (overriding aorta)
   (C)   patent ductus arteriosus
   (D)   ventricular septal defect
   (E)   transposition of the great vessels

354.   This child would be exhibiting what type of blood flow after birth?

   (A)   right-to-left shunt
   (B)   left-to-right shunt
   (C)   normal flow with no shunting
   (D)   alternating right-to-left, then left-to-right shunting

ANSWERS AND TUTORIAL ON ITEMS 352-354

The answers are: **352-C; 353-E; 354-A**. Given just these clinical findings, the child most likely suffers from **tetralogy of Fallot**, the most common cause of cyanosis in a newborn infant. This classic congenital anomaly consists of four defects (1) **pulmonary stenosis**; (2) **ventricular septal defect** (VSD); (3) **overriding aorta** (a rightward displacement of the aorta; and (4) **right ventricular hypertrophy**. It results from unequal bulbotruncal septation such that the pulmonary trunk is stenotic. Additionally, the aorta is enlarged and pulled to the right by the abnormal septation of the bulbotruncal region. Since the bulbotruncal septum also contributes to final closure of the interventricular septum, a ventricular septal defect also results. Because of the stenosis, the right ventricle cannot pump blood through the pulmonary trunk and therefore has to pump its blood through the VSD and into the aorta. Thus, the right ventricle is pumping blood against high resistance which results in right ventricular hypertrophy. Pumping venous blood into the aorta is classified clinically as a **right-to-left shunt** and results in the cyanosis seen in these children.

Items 355-357

The newborn child of a well-known movie star fails to thrive after birth, and on auscultation you hear a distinct heart murmur (abnormal heart sound). Radiological examination is done and you inform the parents that their child has a "hole in her heart" which is the **MOST COMMON** type of congenital heart malformation.

355. The child **MOST** likely has

    (A) tetralogy of Fallot
    (B) primum type atrial septal defect
    (C) ventricular septal defect
    (D) patent ductus arteriosus
    (E) secundum type atrial septal defect

356. You would expect the patient to have which type of blood shunting?

    (A) right-to-left
    (B) left-to-right
    (C) alternating between right-to-left, then left-to-right
    (D) no shunting of blood

357. The parents ask for a simple explanation of the defect and you tell them it **MOST** likely results from a failure of development of

   (A)   septum secundum
   (B)   membranous part of the interventricular septum
   (C)   septum primum
   (D)   ductus arteriosus
   (E)   muscular part of the interventricular septum

ANSWERS AND TUTORIAL ON ITEMS 355-357

The answers are: **355-C; 356-B; 357-B. Ventricular septal defects** (VSDs) are the most common type of congenital heart defect, whether as an isolated defect or in combination with other types of congenital anomalies. The interventricular septum forms as two parts, a **muscular part** derived from the myocardium of the primitive ventricle and a **membranous part** derived from the endocardial cushions and the bulbotruncal septum. Ninety percent of VSD's are a defect in membranous portion of the interventricular septum and result in abnormal left-to-right shunt after birth since left ventricular pressure normally exceeds right ventricular pressure after birth. The clinical picture varies depending on a number of factors such as the patient's age, the size of the defect and the state of vascular development within the lungs. A small defect is usually asymptomatic and sometimes spontaneously resolves after birth. With a medium defect the child may be unable to tolerate physical effort and fatigues easily. With a large defect there is likely to be left atrial and left ventricular hypertrophy. The left ventricle eventually becomes overloaded and may fail.

CHAPTER VII
DEVELOPMENT OF THE
GASTROINTESTINAL SYSTEM

Items 358-359

Choose the **BEST** response.

During the first month of postnatal life, a child is jaundiced and has pale stools. A diagnosis is confirmed during surgical exploration.

358. What is the **MOST** common biliary anomaly of clinical relevance encountered during infancy?

    (A)   biliary atresia
    (B)   floating gallbladder
    (C)   retrodisplacement of the gallbladder
    (D)   agenesis of the gallbladder
    (E)   None of the above.

359. The **MOST** common treatment for this condition is

    (A)   none, it will resolve on its own
    (B)   surgical
    (C)   none, it is always fatal
    (D)   liver transplantation
    (E)   dietary modification

ANSWERS AND TUTORIAL ON ITEMS 358-359

The answers are: **358-A; 359-B**. **Anomalies of the biliary tract** may be found in 10 to 20% of the population. **Biliary atresia** and hypoplasia of the extrahepatic and major intrahepatic bile ducts are the most common biliary anomalies of clinical relevance encountered in infancy. The clinical picture is one of **severe obstructive jaundice** during the first month of postnatal life, with pale stools. The diagnosis is confirmed by surgical exploration with operative cholangiography.

Approximately 10% of biliary atresias are treatable with **choledochojejunostomy** (called the Roux-en-Y procedure). This involves connecting the common bile duct to the jejunum. The remainder of biliary atresias are treated with **hepatic portoenterostomy** (Kasai procedure) in an effort to restore some bile flow. Unfortunately, most children with biliary atresia, even those having successful biliary-enteric anastomoses, eventually develop chronic cholangitis, extensive hepatic fibrosis, and portal hypertension.

Other malformations of the biliary system include abnormalities in number, size and shape of the gallbladder (agenesis of the gallbladder, duplications, rudimentary, or oversized "giant" gallbladder, and diverticula. In addition, anomalies of position or suspension are not uncommon. These include left-sided gallbladder, intrahepatic gallbladder, retrodisplacement of the gallbladder, and "floating" gallbladder. The latter predisposes to torsion, volvulus, or herniation of the gallbladder.

Items 360-365

Choose the **BEST** response.

360.   The liver develops

    (A)   as a diverticulum of the foregut in the region of the presumptive duodenum
    (B)   as an outgrowth from the midgut
    (C)   within the dorsal mesentery
    (D)   from splanchnopleuric mesoderm
    (E)   without association with the septum transversum

361.   The hepatic diverticulum gives rise to all of the following **EXCEPT**:

    (A)   liver parenchymal cells
    (B)   bile canaliculi
    (C)   hepatic duct
    (D)   supporting stroma

362.   Typical bile, secreted by hepatic cells

    (A)   occurs in fetuses 5 months old
    (B)   is excreted by the allantois
    (C)   is produced by the embryonic spleen
    (D)   All of the above.
    (E)   Only A and C above.

363.   The endodermal hepatic diverticulum differentiates into all of the following **EXCEPT**:

    (A)   ventral part of the pancreas
    (B)   gallbladder
    (C)   spleen
    (D)   common bile duct
    (E)   cystic duct

364.   The hepatic diverticulum grows into the

    (A)   dorsal mesentery
    (B)   dorsal mesogastrium
    (C)   septum transversum
    (D)   pleuroperitoneal canals
    (E)   splanchnic mesoderm

365.   Which of the following events in the development of the abdominal cavity is affected by the rapid growth of the liver?

    (A)   urorectal septum formation
    (B)   formation of the greater omentum
    (C)   formation of the inferior recess of the lesser sac
    (D)   herniation of the midgut loop
    (E)   descent of the gonads

ANSWERS AND TUTORIAL ON ITEMS 360-365

The answers are: **360-A; 361-D; 362-A; 363-C; 364-C; 365-D**. The liver begins development on about day 22 as a diverticulum (**hepatic diverticulum**) which grows into the ventral region of the **septum transversum**. The hepatic diverticulum gives rise to the **liver cords**, which become the hepatocytes (parenchyma), to the **bile canaliculi** of the liver, and to the **hepatic ducts**. The mesoblastic supporting stroma of the liver develops from splanchnopleuric mesoderm of the septum transversum. The **endoderm** that forms the hepatic diverticulum ultimately differentiates not only into the parenchyma of the liver, but also into the gallbladder and the ventral pancreatic bud. The **cystic diverticulum** which buds from the hepatic diverticulum, will form the **gallbladder** and the **cystic duct**.

    The hepatic cells of the liver begin to secrete bile by the time the fetus is five months old. The mass of mucus, bile and desquamated epithelial cells, plus swallowed secretions, constitute the **meconium**. The bile, which is produced by the liver cells in the five month-old fetus, gives the meconium its typical dark-green coloration.

The **spleen** arises as a condensation of mesenchymal cells within the dorsal mesogastrium. Therefore, the spleen is a mesodermal derivative, not a product of the gut tube endoderm.

The rapid growth of the liver has considerable effect on the development of the ventral mesentery, and it promotes the **herniation of the midgut** by occupying space in the abdominal cavity. It does not affect urorectal septum formation, formation of the greater omentum, formation of the inferior recess of the lesser sac, or descent of the gonads.

Items 366-368

Choose the **BEST** response.

366.   All of the following are correct concerning the ventral mesentery **EXCEPT**:

    (A)   gives rise to the falciform ligament
    (B)   gives rise to the greater omentum
    (C)   portion becomes the hepatoduodenal ligament
    (D)   portion becomes the hepatogastric ligament
    (E)   portion becomes the lesser omentum

367.   In the fetus, the caudal, free margin of the falciform ligament contains the

    (A)   umbilical artery
    (B)   sinus venosus
    (C)   ductus venosus
    (D)   common bile duct
    (E)   umbilical vein

368.   The epiploic foramen is an opening between the

    (A)   omental bursa and greater peritoneal cavity
    (B)   peritoneal cavity
    (C)   thoracic and abdominal cavities
    (D)   abdominal and pelvic cavities
    (E)   peritoneal and pericardial cavities

ANSWERS AND TUTORIAL ON ITEMS 366-368

The answers are: **366-B; 367-E; 368-A**. The **hepatic diverticulum** grows into the ventral mesentery and the septum transversum. The ventral mesentery is modified to form a number of membranous structures. It will give rise to the serous covering of the liver and the peritoneal folds that attach the liver to the stomach and to the ventral body wall. The narrow, sickle-shaped segment of ventral mesentery that attaches the liver to the ventral body wall differentiates into the **falciform ligament**. The caudal, free margin of the falciform ligament carries the umbilical vein from the body wall to the liver. After birth, this vein degenerates and forms the **ligamentum teres hepatis** (round ligament of the liver). The portion of the ventral mesentery between the liver and the stomach thins out to form the **lesser omentum**. The portion of the lesser omentum, connecting the liver to the developing duodenum, is call the **hepatoduodenal ligament** and contains the hepatic portal vein, the hepatic artery and the common bile duct (hepatic triad). The portion of the lesser omentum between the liver and the stomach is called the **hepatogastric ligament**.

When the stomach rotates to the left and the liver shifts its position to the upper right quadrant of the abdominal cavity, the lesser omentum also rotates from a sagittal plane to a coronal plane. Thus, the communication or opening between the greater and lesser peritoneal sacs is reduced to a narrow canal called the **epiploic foramen (of Winslow)**. The **epiploic foramen** is the only passageway between the lesser peritoneal sac (omental bursa) and the greater peritoneal cavity. In the adult, it serves as a landmark for the biliary structures and inferior vena cava during gallbladder and other types of abdominal surgery. The triad of biliary structures is anterior to the foramen and the inferior vena cava is posterior.

<u>Items 369-374</u>

Choose the **BEST** response.

369.   Which of the following characterizes early development of the midgut?

    (A)   formation of the primary intestinal loop
    (B)   elongation
    (C)   Both A and B above.
    (D)   Neither A nor B above.

370.   Factors, major events, or structures associated with midgut development include

    (A)   270° total rotation
    (B)   rapid growth of cranial limb
    (C)   physiological hernia
    (D)   All of the above.
    (E)   None of the above.

371. The axis for rotation of the midgut loop is the

    (A)   cecum
    (B)   duodenum
    (C)   aorta
    (D)   superior mesenteric artery
    (E)   celiac artery

372. The vermiform appendix arises from

    (A)   endodermal outgrowth of the cecum
    (B)   splanchnic mesoderm
    (C)   Both A and B above.
    (D)   Neither A nor B above.

373. Retraction of the midgut hernia

    (A)   occurs during the 10th week of development
    (B)   involves a further 180° rotation
    (C)   occurs as a result of poorly understood factors
    (D)   All of the above.
    (E)   None of the above.

374. All of the following portions of the gastrointestinal tract are classified as being intraperitoneal **EXCEPT**:

    (A)   transverse colon
    (B)   sigmoid colon
    (C)   most of the duodenum
    (D)   ileum
    (E)   jejunum

ANSWERS AND TUTORIAL ON ITEMS 369-374

The answers are: **369-C; 370-D; 371-D; 372-C; 373-D; 374-C**. By the fifth week of development, the presumptive ileum begins to elongate rapidly. The ileum lengthens more rapidly than the abdominal cavity enlarges, and the midgut becomes a fold called the **midgut loop** or **primary intestinal loop**. The cranial limb of the loop will give rise to all of the jejunum and most of the ileum. The caudal limb will give rise to the ascending and transverse colon. The apex of the loop is attached to the umbilicus by the vitelline duct. The superior mesenteric artery travels through the dorsal mesentery to the apex of the midgut loop. By the 6th week the continuing elongation of the midgut loop, combined with the dynamic growth of other abdominal organs such as the liver and kidneys, forces the midgut loop to leave the abdominal cavity and herniate into the umbilicus. This is known as a **physiological hernia** since it is considered a normal part of intestinal development.

As the midgut loop herniates into the umbilicus, it rotates around the axis of the **superior mesenteric artery**. While in the umbilicus, the midgut **rotates 90° counterclockwise** such that the cranial limb moves caudally and the caudal limb moves cranially to the embryo's left. This rotation is complete by the early part of the 8th week. During this time, the midgut continues to differentiate, with further lengthening of the jejunum and ileum forming the jejunal-ileal loops. The cecum expands at this time and sprouts the **vermiform appendix**. Like the rest of the gut, it has an endodermally-lined lumen that has a muscular wall derived from the **splanchnic mesoderm**.

During the **10th week** of development, the midgut loop hernia is rapidly retracted into the abdominal cavity. The factors which are responsible for this retraction are not well understood, but may involve an increase in the size of the abdominal cavity relative to the other abdominal organs. As the intestinal loop re-enters the abdominal activity it undergoes a further **180° counterclockwise rotation**. Thus, the midgut rotates a total of 270° counterclockwise. The caudal limb of the midgut is the last to be retracted and consequently the cecum comes to lie just inferior to the liver. By the 11th week the intestines have completely returned to the abdominal cavity.

After development, herniation and rotation of the midgut, the jejunum and ileum retain their dorsal mesentery and thus are classified as being **intraperitoneal organs**. The dorsal mesenteries of the ascending and descending colon shorten and disappear, bringing these organs into contact with the dorsal body wall. The organs adhere to the wall and thus become **secondarily retroperitoneal**. The transverse colon and sigmoid colon do not become fixed to the body wall, retain there mesenteries and are thus classified as being **intraperitoneal**. Pressure from the transverse colon may help to fix the underlying duodenum which becomes mostly a retroperitoneal organ secondarily.

Items 375-377

Choose the **BEST** response.

375.   All of the following statements about development of the stomach are correct
       EXCEPT:

    (A)   It undergoes rotation around both its longitudinal and dorsoventral axes.
    (B)   The original dorsal surface becomes the greater curvature.
    (C)   The left vagus nerve comes to lie on the anterior surface
    (D)   The original right surface becomes the lesser curvature.
    (E)   The original left surface becomes the anterior surface.

376. Rotation of the stomach occurs during development because of

    (A)   rapid expansion of the dorsal mesentery
    (B)   slow growth of the ventral mesentery
    (C)   Both A and B above.
    (D)   Neither A nor B above.

377. A consequence of rotation of the stomach is

    (A)   a space called the lesser sac forms
    (B)   the duodenum and pancreas become retroperitoneal
    (C)   the dorsal mesogastrium becomes the greater omentum
    (D)   All of the above.
    (E)   None of the above.

ANSWERS AND TUTORIAL ON ITEMS 375-377

The answers are: **375-D; 376-C; 377-D**. During the 7th and 8th weeks the **stomach** rotates 90° around its longitudinal axis and also rotates around its dorsoventral axis so that the original dorsal surface is directed to the left side and slightly caudally and becomes the greater curvature. This rotation shifts the liver to the right in the abdominal cavity and also brings the duodenum and pancreas into contact with the posterior body wall, where they become fixed (i.e., secondarily retroperitoneal). The peritoneal space to the right of the unrotated stomach and the dorsal mesogastrium are converted through rotation into a space called the **lesser sac of the peritoneum**. Due to voluminous expansion, the dorsal mesogastrium extends downward over the transverse colon and becomes the curtain-like **greater omentum**. Rotation of the stomach also moves the left and right vagal plexuses which originally ran along the right and left sides of the gut tube. Thus, the left vagus nerves becomes the anterior vagal trunk (anterior gastric nerve) and the right vagus becomes the posterior vagal trunk.

Choose the **BEST** response.

378.   The pancreas arises embryologically

      (A)   as a single diverticulum of the foregut
      (B)   spontaneously within the septum transversum
      (C)   as a result of migration of neural crest cells
      (D)   as two diverticula of the foregut
      (E)   as a collection of mesenchymal tissue in the dorsal mesentery

379.   The main pancreatic duct of the adult pancreas

      (A)   is formed by fusion of the dorsal duct with the ventral duct
      (B)   is derived entirely from the common bile duct
      (C)   is derived entirely from the dorsal pancreatic duct
      (D)   empties into the minor papilla of the duodenum

380.   The head of the pancreas arises from the

      (A)   ventral pancreatic bud alone
      (B)   dorsal pancreatic bud alone
      (C)   common bile duct
      (D)   gallbladder
      (E)   dorsal and ventral pancreatic bud

381.   An accessory pancreas

      (A)   is always retroperitoneal
      (B)   may occur within the wall of the intestine and stomach
      (C)   develops within the liver
      (D)   wraps around the duodenum forming an annular pancreas

ANSWERS AND TUTORIAL ON ITEMS 378-381

The answers are: **378-D; 379-A; 380-E; 381-B**. The **pancreas** begins on day 26 as dorsal and ventral buds of endoderm. The dorsal bud arises as a diverticulum of the foregut (presumptive second part of the duodenum). A few days later another diverticulum, the ventral pancreatic bud, sprouts from the developing biliary diverticulum (presumptive common bile duct). The distal end of the common bile duct and the ventral pancreatic bud move dorsally around the duodenum to the dorsal mesentery. By the beginning of the 6th week the ventral and dorsal pancreatic buds lie adjacent do one another in the dorsal mesentery and by the end of the 6th week they fuse to form the definitive pancreas. The dorsal pancreatic bud gives rise to the **cranial part of the head**, the **body**, and the **tail** of the pancreas while the ventral pancreatic bud gives rise to the caudal part of the head and the **uncinate process**. The proximal part of the duct from the dorsal bud to the duodenum usually degenerates, leaving the ventral pancreatic duct as the main pancreatic duct. The main duct joins with the common bile duct to from the **ampulla of Vater**. The ampulla empties into the duodenum at the **major duodenal papilla**. The dorsal pancreatic duct may persist as an accessory pancreatic duct that empties into the duodenum at the minor duodenal papilla.

      Accessory or **heterotopic pancreatic tissue** can occur in the wall of the stomach and intestine, in Meckel's diverticulum, and sometimes it is associated with the spleen. An **annular pancreas** forms a complete ring encircling the duodenum. This abnormality probably arises when the ventral pancreatic bud moves in the opposite direction around the duodenum and fuses with the dorsal pancreatic bud. An annular pancreas may compress the duodenum and cause gastrointestinal obstruction.

<u>Items 382-388</u>

Choose the **BEST** response.

382.   The embryonic foregut differentiates into all or part of the

    (A)   liver
    (B)   ventral pancreas
    (C)   esophagus
    (D)   lung
    (E)   All of the above.

383.   Which of the following portions of the early gut tube is located immediately dorsal to the yolk sac?

    (A)   foregut
    (B)   midgut
    (C)   hindgut
    (D)   All of the above.
    (E)   None of the above.

384. Which of the following parts of the developing gut is correctly matched with its arterial supply?

(A)   midgut - superior mesenteric artery
(B)   foregut - celiac artery
(C)   hindgut - inferior mesenteric artery
(D)   All of the above are correctly matched.
(E)   None of the above are correctly matched.

385. The foregut terminates cranially at the

(A)   thyroid diverticulum
(B)   septum transversum
(C)   oropharyngeal membrane
(D)   laryngotracheal groove
(E)   pharyngeal pouches

386. Meckel's diverticulum is a remnant of the

(A)   urachus
(B)   umbilical cord
(C)   yolk sac
(D)   vitelline duct
(E)   allantois

387. The clinical significance of a Meckel's diverticulum is

(A)   none, it usually regresses with no consequence
(B)   it may become obstructed and inflamed
(C)   it may result in volvulus
(D)   it may result in gastroschisis
(E)   it may result in hiatal hernia

388. Persistence of the vitelline duct may also result in

(A)   vitelline cyst
(B)   umbilical fistula
(C)   Both A and B above.
(D)   Neither A nor B above.

ANSWERS AND TUTORIAL ON ITEMS 382-388

The answers are: **382-E; 383-B; 384-D; 385-C; 386-D; 387-B; 388-C**. When the embryo undergoes folding, the gut consists of a cranial and caudal blind-ending tube, the presumptive **foregut** and **hindgut**, and a central **midgut** , which still opens ventrally to the yolk sac. The foregut terminates cranially at the **oropharyngeal membrane**; and, caudally, the hindgut terminates at the **cloacal membrane**. As the gut lengthens and the embryo enlarges and continues to fold, the open portion of the midgut narrows until it becomes the slender vitelline duct.

Derivatives of embryonic **foregut** include the pharynx and pahryngeal pouches, esophagus, lungs and respiratory system, thyroid gland, pharyngeal pouches, stomach, the duodenum cranial to the hepatic diverticulum (1st and 2nd parts of the duodenum in the adult), the pancreas, liver, gall bladder, and biliary tree.

The boundaries of the foregut, midgut and hindgut correspond to the territories of the three arteries that supply the abdominal portion of the gut tube. The **celiac trunk** supplies the abdominal part of the foregut, (abdominal esophagus, stomach, and superior half of the duodenum and its derivatives). The **superior mesenteric artery** supplies the midgut. The **inferior mesenteric artery** supplies the hindgut.

The **vitelline duct** is a connection between the embryonic midgut and the yolk sac. As development proceeds and placental nutrition becomes established, the vitelline duct normally regresses by the end of the 7th week. Persistence of the duct gives rise to a number of abnormalities, the most common is **Meckel's diverticulum**, a 1-2 inch long intestinal pouch resulting from the persistence of the enteric (intra-abdominal) portion of the vitelline duct. Meckel's diverticulum occurs in about 2% of the adult population and is found along the distal 30 inches of the ileum. Meckel's diverticulum is described by surgeons as the "disease of two": affects 2% of population, lies 2 feet from the ileocecal valve, is 2 inches long, occurs in men 2 times as often as women, contains 2 types of ectopic tissue (gastric and/or pancreatic) and has 2 major complications (hemorrhage and inflammation).

Persistence of the vitelline duct may also give rise to a **vitelline duct cyst**, an **umbilical/vitelline fistula**, or a fibrous cord connecting the gut to the umbilicus. A fistula involving the umbilicus may involve either the allantoic or vitelline duct. If the vitelline duct remains patent, there will be an umbilical or vitelline fistula, leaking **meconium** or fetal feces at the umbilicus. Partial closure of the vitelline duct may result in a vitelline cyst.

Items 389-392

Choose the **BEST** response.

389. Which one of the following anomalies is **MOST** common in live-born males?

(A)   pyloric stenosis
(B)   dislocated hip
(C)   birthmarks
(D)   brain defects
(E)   congenital inguinal hernia

390. Which of the following conditions is **MOST** likely to cause vomiting of bile-stained fluid in the newborn?

(A)   umbilical fistula
(B)   duodenal stenosis
(C)   rectal atresia
(D)   biliary atresia
(E)   esophageal stenosis

391. The **MOST** frequent cause of intestinal obstruction in the newborn is

(A)   intestinal atresia
(B)   nonrotation
(C)   Meckel's diverticulum
(D)   omphalocele
(E)   gastroschisis

392. Intestinal obstruction such as stenosis or atresia is **MOST** often caused by

(A)   physiological hernia
(B)   failure of recanalization of the epithelial plug
(C)   gastroschisis
(D)   All of the above.
(E)   None of the above.

ANSWERS AND TUTORIAL ON ITEMS 389-392

The answers are: **389-A; 390-B; 391-A; 392-B**. During the 6th week the endodermal epithelium of the gut tube proliferates until it completely occludes the lumen (formation of an **epithelial plug**). During the next 2 weeks vacuoles develop in the tissue and coalesce until the gut tube is fully recanalized. During the 9th week the definitive mucosal epithelium differentiates from the endodermal lining of the new gut lumen.

Certain anomalies are found to be more prevalent in one sex than the other. For example, birthmarks, anomalies of brain and spinal cord, and dislocated hip are more common in females, while cleft lip and palate, **pyloric stenosis** and hydrocephalus are more common in males. Congenital hypertrophic pyloric stenosis is an abnormality of the pyloric musculature which results in mechanical obstruction of the distal portion of the stomach. There is an increase in the number smooth muscle fibers of the pylorus resulting in a fusiform mass. This leads to excessive and continued vomiting with consequent loss of fluid and electrolytes. Pyloric stenosis occurs most frequently in firstborn male children with a ratio of 4:1 male:female. The initial symptom is usually forceful (projectile) vomiting of the gastric contents. A firm, olive-shaped mass sometimes can be palpated in left upper quadrant. The treatment usually is surgical.

Vomiting in the newborn is most likely to be caused by an obstruction in the upper part of the GI tract. **Esophageal atresia** or duodenal stenosis could cause this.

Congenital interruption of the alimentary tract may occur anywhere between the pharynx and anus. **Intestinal atresia** represents the most frequent cause of intestinal obstruction in the newborn. The ileum is affected in approximately half the cases with the duodenum and jejunum following in frequency. Multiple areas of atresia may be present throughout a segment of the small intestine. The cause of intestinal atresia is associated with the proliferative stage of the epithelium during normal intestinal development. Epithelial proliferation causes obliteration of the lumen by formation of a plug that is later recanalized. If recanalization does not re-establish the intestinal lumen, atresia results. Another factor which has been implicated in intestinal atresia is vascular injury resulting from intussusception, volvulus, or strangulation during the period of physiologic herniation of the small intestine into the umbilical stalk. The high incidence of volvulus in association with atresia lends support to this theory.

Signs and symptoms of intestinal obstruction are dependent on the level of the obstruction. Characteristically, vomiting occurs shortly after birth but it may be delayed for 24 hours with low obstruction. Vomiting of bilious material is suggestive particularly of mechanical intestinal obstruction. **Polyhydramnios** should alert the physician to the possibility of intestinal obstruction. An abnormally large amount of amniotic fluid may be caused by interference with the normal passage of swallowed amniotic fluid into the small intestine where it is absorbed. X-ray reveals the accumulation of gas proximal to the point of obstruction.

Items 393-395

Choose the **BEST** response.

393. Which of the following types of hernia is considered normal in the course of embryological development?

    (A)   inguinal
    (B)   umbilical
    (C)   lumbar
    (D)   diaphragmatic
    (E)   femoral

394. Failure of the retraction of the midgut loop may result in a condition known as

    (A)   ectopia cordis
    (B)   stenosis
    (C)   gastroschisis
    (D)   omphalocoel
    (E)   prune belly

395. An example of a consequences of abnormal rotation of the midgut loop is

    (A)   Meckel's diverticulum
    (B)   omphalocoele
    (C)   duodenal stenosis
    (D)   esophageal fistula
    (E)   volvulus

ANSWERS AND TUTORIAL ON ITEMS 393-395

The answers are: **393-B; 394-D; 395-E**. The **herniation of the midgut loop** into the proximal part of the umbilical cord is a normal event during development. It is referred to as a **physiological hernia**. The herniated intestines may fail to return to the abdomen, resulting in an **omphalocele**. Viscera may herniate postnatally through the weak umbilical region (i.e., through the umbilical ring), resulting in a true umbilical hernia. Omphalocele occurs in about 2.5 per 10,000 births. An omphalocele may occur in conjunction with a variety of cardiac and renal defects, often as part of a constellation of abnormalities associated with a chromosomal anomaly. One example is the **pentalogy of Cantrell**, which includes an omphalocoele, diaphragmatic hernia, sternal cleft, ectopia cordis (ectopic heart) and intracardiac anomalies.

Rotational defects of the midgut can be classified as nonrotations, reversed rotations, and mixed rotations/malrotations. During the sixth to tenth week of embryonic life a portion of the midgut protrudes into the proximal part of the umbilical cord. When the abdominal cavity achieves a size which can accommodate the intestine, the midgut returns to the abdomen and rotates 180° in a counterclockwise direction. Subhepatic cecum and volvulus are two lesions which result from incomplete or abnormal rotation of the midgut loop. Incomplete rotation of the cecum about the superior mesenteric vessels causes it comes to remain in the right upper quadrant. The malrotated cecum is usually fixed in the right upper quadrant by peritoneal reflections. These peritoneal reflections form **Ladd bands** that frequently overlie the distal portion of the duodenum and result in extrinsic compression.

Symptoms of **duodenal compression** usually consist of vomiting of bile-stained material. Prolonged vomiting results in dehydration and electrolyte depletion. X-ray studies reveal a gas filled and dilated stomach and proximal duodenum. Barium study reveals the duodenal obstruction and the barium enema shows the cecum in an incompletely rotated position. Malrotation of the intestines is the second most common cause of neonatal duodenal obstruction, after the more common duodenal atresia.

Midgut volvulus is commonly associated with incomplete rotation of the midgut loop and results from a failure of fixation of the posterior mesentery and a consequence twisting of the intestines, from the duodenal-jejunal junction to the midtransverse colon. Volvulus occurs usually in a clockwise direction around the proximal segment of the mesenteric vessels. Torsion may partially or completely occlude the superior mesenteric vessels producing ischemic necrosis in addition to obstruction.

Items 396-397

Choose the **BEST** response.

396.   The terminal, dilated part of the hindgut is called the

    (A)   cloaca
    (B)   tail gut
    (C)   transverse colon
    (D)   cecum
    (E)   sigmoid colon

397. The pectinate line in the wall of the anal canal marks the junction of

    (A)   skin and mucous membrane
    (B)   splanchnic and somatic mesoderm
    (C)   ectoderm and endodermal
    (D)   All of the above.
    (E)   None of the above.

ANSWERS AND TUTORIAL ON ITEMS 396-397

The answers are: **396-A; 397-D**. The folding of the tail region of the embryo brings together the allantois and the hindgut. The dilated, terminal part of the hindgut is called the **cloaca** which communicates with the allantois. The allantois is a diverticulum of the cloaca that extends into the connecting stalk. Later in development the cloaca is divided by the **urorectal septum** into the primitive **urogenital sinus** and the **anorectal canal**. The urogenital sinus will give rise to the bladder, the pelvic urethra, and the penile urethra.

The superior two-thirds of the anal canal forms from the distal part of the hindgut. The inferior one-third is derived from an ectodermal pit called the **anal pit** or **proctodeum**. The anal membrane which separates the endodermal and ectodermal porions of the anal canal breaks down in the 8th week. The embryonic location of the membrane in the adult is marked by an irregular folding of mucosa within the anal canal called the **pectinate line**. Early in development the cloacal membrane is a fusion of ectoderm and endoderm and separates the amniotic cavity from the yolk sac cavity. Yolk sac endoderm forms the mucosa of the gut, and the walls of the gut develop from splanchnic mesoderm. Thus, the pectinate line marks the junction of **skin and mucous membrane, splanchnic and somatic mesoderm** and **ectoderm and endoderm**.

Examine **Figure 7.1** below and then match the **BEST** descriptive statement about the development with the appropriately lettered adult structure. Answers may be used once, more than once, or not at all.

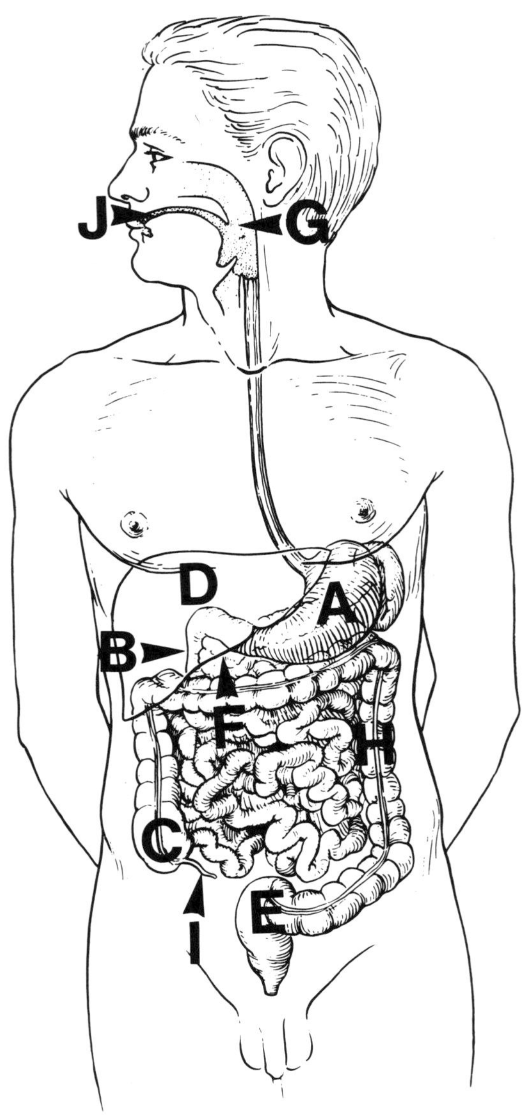

Figure 7.1

398. This structure derives its connective tissue and blood vessels from the mesenchyme of the septum transversum.

399. This structure forms from two separate embryonic structures that fuse during rotation of the midgut loop.

400. The dorsal mesentery of this structure grows extensively to form the greater omentum.

401. This structure is a hindgut derivative that becomes secondarily retroperitoneal.

402. Epithelial lining is derived from stomodeal ectoderm.

ANSWERS AND TUTORIAL ON ITEMS 398-402

The answers are: **398-D; 399-F; 400-A; 401-H; 402-J. Figure 7.1** is a diagram of the adult gastrointestinal tract. The **oral cavity** (J) develops from an ectodermal depression in the face region called the stomodeum. The stomodeum is separated from the cranial end of the foregut for a brief period by the oropharyngeal membrane. The **foregut** derivatives include everything from the **oropharyngeal membrane** to the major papilla where the common bile duct enters the duodenum. Thus, it forms the lining of the pharynx (G), the esophagus, the stomach, and the upper duodenum. During its complex rotation, the dorsal mesentery of the **stomach** (A) grows extensively to form the **greater omentum**. Since the caudal part of the foregut is supplied by the celiac artery, its derivatives (i.e., stomach, duodenum, liver, and pancreas) are also supplied by this artery.

The **midgut** (supplied by the superior mesenteric artery) forms the lower portion of the **duodenum** (B) caudal to the major papilla, the rest of the small intestines, and the large intestines up to the middle of the transverse colon. The caudal segment of the **duodenum** (B) is a derivative of the midgut loop and, therefore, is supplied by the superior mesenteric artery. The jejunum, ileum, **cecum** (C), **appendix** (I), ascending colon, and first part of the transverse colon are also midgut derivatives. The rest of the transverse colon, the **descending colon** (H), the **sigmoid colon** (E), the rectum, and the upper part of the anal canal are all **hindgut** derivatives and are supplied by the inferior mesenteric artery. The ascending (midgut) and descending (hindgut) colon develop as intraperitoneal organs but their dorsal mesenteries fuse with the dorsal body wall so that they become **secondarily retroperitoneal**. The transverse colon remains as an intraperitoneal organ.

The **liver** (D) forms from the hepatic diverticulum which grows into the nearby splanchnic mesoderm of the **septum transversum**. The hepatic diverticulum forms anastomosing epithelial cords of hepatic parenchymal cells as well as the hepatic bile ducts. The splanchnic mesoderm forms connective tissue, blood vessels (including sinusoidal endothelium), and Kupffer cells. The dorsal and ventral pancreatic buds fuse to form a single **pancreas** (F) during rotation of the stomach.

CHAPTER VIII
DEVELOPMENT OF THE
UROGENITAL SYSTEM

Items 403-410

Choose the **BEST** response

403.   Which of the following adult structures is derived from the ureteric bud?

    (A)   renal corpuscle
    (B)   glomerulus
    (C)   nephron
    (D)   loop of Henle
    (E)   minor calyx

404.   All of the following structures develop from the ureteric bud (metanephric diverticulum) **EXCEPT**:

    (A)   renal pelvis
    (B)   Bowman's capsule
    (C)   ureter
    (D)   collecting tubules
    (E)   major calyces

405.   Which of the following structures develops from the vesical portion of the urogenital sinus?

    (A)   major calyx
    (B)   spongy urethra
    (C)   urinary bladder
    (D)   membranous urethra
    (E)   bulbourethral gland

406.   The cloaca is divided into the rectum and urogenital sinus by the

    (A)   cloacal membrane
    (B)   urogenital ridge
    (C)   sinovaginal bulbs
    (D)   urorectal septum
    (E)   rectal columns

407.   All of the following statements about the metanephros are true **EXCEPT**:

    (A)   It develops from mesoderm only.
    (B)   It is the name of the permanent kidney.
    (C)   It forms from a single primordium.
    (D)   It attains its definitive position during the first trimester.
    (E)   It contains the definitive number of nephrons at full term birth.

408.   All of the following statements about development of the urinary bladder are true
EXCEPT:

    (A)   In the fetus and infant it is located in the abdomen rather than the pelvis.
    (B)   The apex of the fetal urinary bladder is continuous with the umbilicus by way
        of the urachus.
    (C)   It is a pelvic organ in children.
    (D)   It develops mainly from the cranial part of the urogenital sinus.
    (E)   Its epithelial lining is derived mainly from endoderm.

409.   Choose the **TRUE** statement concerning the partitioning of the cloaca.

    (A)   The urorectal septum divides the cloaca into a cranial, vesical part and a
        caudal, pelvic part.
    (B)   The area of fusion of the urorectal septum with the cloacal membrane is
        represented in the adult by the perineal body.
    (C)   The urorectal septum divides the cloacal membrane into a dorsal, hymen and a
        ventral, urethral plate.
    (D)   The urorectal septum develops between the hindgut and the tailgut.
    (E)   The urorectal septum becomes the urogenital membrane.

410.   The mesonephric duct initially empties into

    (A)   ureteric bud (metanephric diverticulum)
    (B)   ventral part of the cloaca
    (C)   urachus
    (D)   pelvic part of the urogenital sinus
    (E)   dorsal part of the cloaca

The answers are: **403-E; 404-B; 405-C; 406-D; 407-C; 408-C; 409-B; 410-B**. The **ureteric bud** or metanephric diverticulum is an outgrowth of the **mesonephric duct** near its entry into the cloaca. The ureteric bud gives rise to the collecting part of the urinary system, i.e., ureter, renal pelvis, calices and collecting tubules.

Bowman's capsule develops from the proximal end of a metanephric tubule that forms from a metanephric vesicle which, in turn, develops from a mass of mesoderm called the **metanephrogenic blastema**.

The vesical part of the **urogenital sinus** gives rise to all of the urinary bladder lining except the trigone area. The prostatic and membranous urethra arises from the pelvic part. Most of the spongy urethra arises from the phallic part.

The **urorectal septum** elongates as the caudal end of the embryo unrolls thereby joining the urorectal septum with the cloaca membrane and separating the rectum/anal canal from the urogenital sinus.

The permanent kidney or **metanephros** develops from two primordia - the **ureteric bud** (metanephric diverticulum) and the **metanephric blastema**. Both primordia originate in the intermediate mesoderm which lies between medial somitic mesoderm and lateral plate mesoderm. The definitive position of the kidney is attained by nine weeks and all nephrons are formed prior to birth, although there is considerable postnatal renal development due to increases in length of tubules and complexity of blood vessels.

The **urinary bladder** does not become a pelvic organ until after puberty. The **cloaca** is divided into a rectum/anal canal and a urogenital sinus by the urorectal septum. The septum fuses with the cloacal membrane and divides it into an anal membrane and a urogenital membrane. The area of fusion of the septum and the membrane is the perineal body in the adult. When the mesonephric duct does not empty into the ventral part of the cloaca the ureter can have an ectopic opening since it forms as a diverticulum of the caudal part of the duct.

Items 411-415

Choose the **BEST** response.

411. Failure of the ureteric bud to form results in

   (A)   renal agenesis
   (B)   urachal sinus
   (C)   pelvic kidney
   (D)   polycystic kidney
   (E)   exstrophy of the urinary bladder

412. Which of the following is the **MOST** common variation/malformation observed in the development of the urinary system?

   (A)   exstrophy of the urinary bladder
   (B)   pancake kidney
   (C)   multiple renal arteries
   (D)   horseshoe kidney
   (E)   congenital polycystic kidney

Figure 8.1 is an anterior view of a horseshoe kidney.

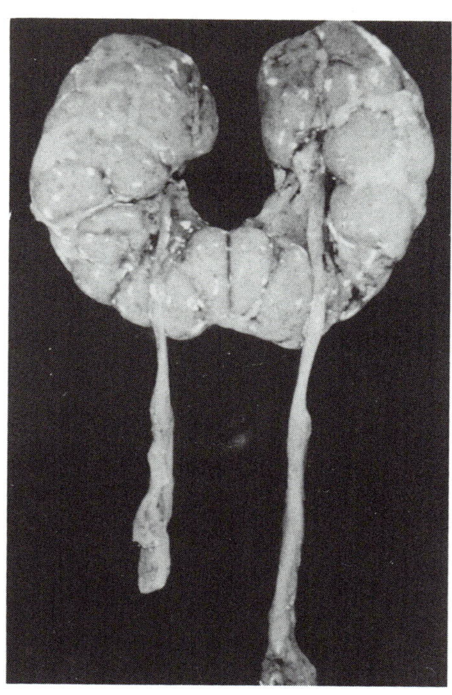

Figure 8.1

413. All of the following statements about horseshoe kidney are true **EXCEPT**:

(A)   It occurs in approximately one in 500 persons.
(B)   It usually produces symptoms.
(C)   Wilms' tumors occur more frequently in children with this condition.
(D)   It usually is located at the level of lower lumbar vertebrae.
(E)   The ureters usually empty into the urinary bladder.

414. When a horseshoe kidney develops as a result of fusion of the inferior poles, the definitive position of the kidney is dictated by the

(A)   position of the adrenal gland
(B)   size of the renal pelvis
(C)   number of collecting tubules within the kidney
(D)   inferior mesenteric artery
(E)   position of the renal arteries

415. All of the following statement about urinary tract malformations/variations are true **EXCEPT**:

(A)   Urinary tract duplications are common.
(B)   Urinary tract duplications result from premature division of the metanephric diverticulum.
(C)   When present, ectopic ureteric orifices often open into the neck of the urinary bladder.
(D)   Ureteric ectopia occurs when the mesonephric duct is incorporated into the trigone of the urinary bladder.
(E)   Incontinence is the common complaint from ureteric ectopia.

ANSWERS AND TUTORIAL ON ITEMS 411-415

The answers are: **411-A; 412-C; 413-B; 414-D; 415-D**. **Renal agenesis** (absence of the kidney, either unilateral or bilateral) occurs when there is a failure of the reciprocal inductive epithelial mesenchymal interaction between the ureteric bud and the metanephric blastema. When the **ureteric bud** fails to penetrate the metanephric blastema, it fails to induce nephron formation and does not become extensively branched. Approximately 30% of people have more than one renal artery. This common variation reflects the manner in which the blood supply to the kidney continually undergo changes during its formation.

Horseshoe kidney is a relatively frequently occurring condition that usually produces no symptoms. It is usually located in the lower lumbar region but its short ureters usually empty into the urinary bladder. A rapidly forming malignant tumor called Wilms' tumor occurs about five times more frequently in children with horseshoe kidney. The horseshoe shape usually is caused by fusion of the inferior poles in which case complete "ascent" of the kidneys is prevented by the root of the inferior mesenteric artery that lies immediately superior to the fused area.

    Urinary tract duplications are common and are caused by premature splitting of the ureteric bud. An **ectopic ureteric orifice** occurs when the mesonephric duct is **NOT** incorporated into the trigone but is incorporated instead into the dorsal part of the cloaca or the caudal part of the urogenital sinus. Usual sites for the ectopic orifice is the urinary bladder neck and the prostatic urethra. Incontinence occurs in these cases because the ureters do not open into the urinary bladder but into the urethra or urinary bladder neck instead, downstream from the sphincter of the urinary bladder.

<u>**Items 416–421**</u>

A newborn infant was delivered normally but died shortly after birth. At autopsy, the kidneys shown in **Figure 8.2** were observed. A suspected diagnosis of congenital polycystic kidney was confirmed when histologic pathology was completed.

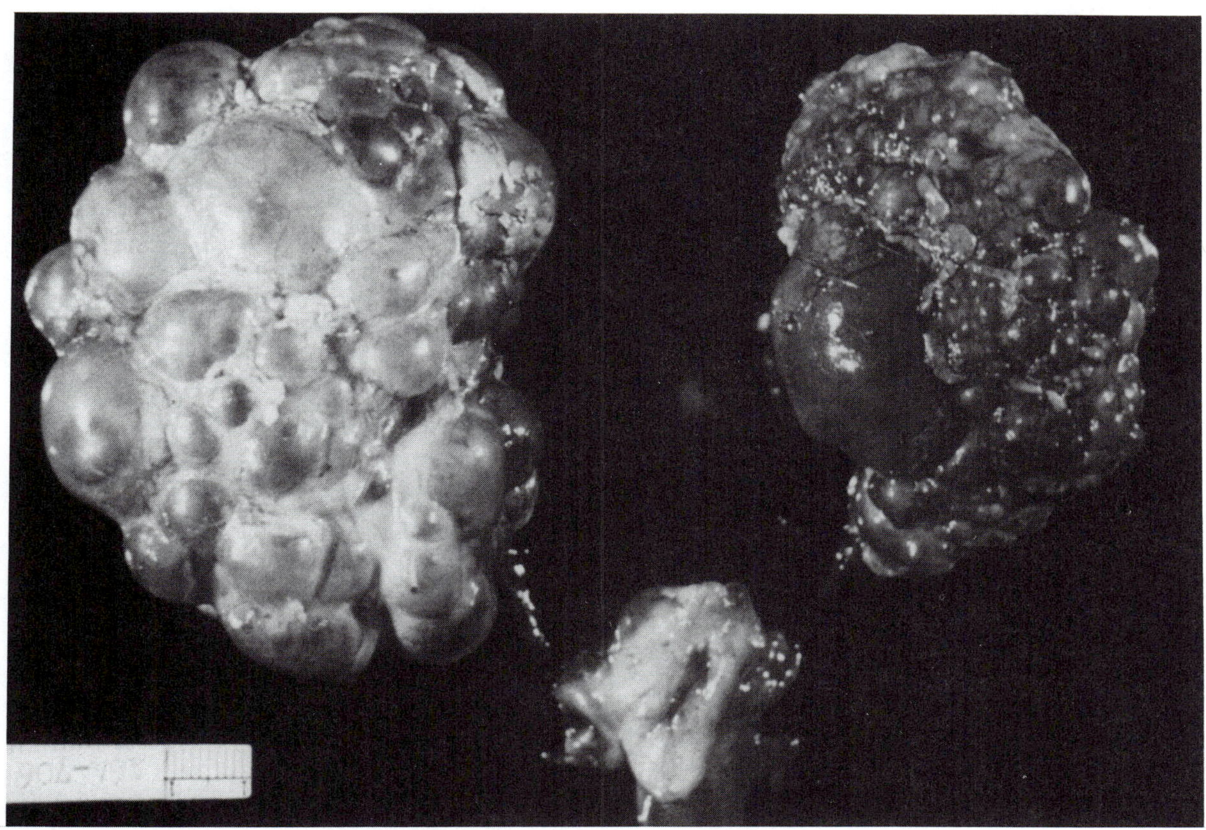

Figure 8.2

Choose the **BEST** response.

416.   Which of the following statements is **TRUE**?

    (A)   Congenital polycystic kidney disease is relatively uncommon.
    (B)   Death is usually caused by the pressure of the cysts on surrounding organs.
    (C)   The cysts are wide dilations of parts of nephrons.
    (D)   Kidney transplant is of little benefit to such infants because the transplanted kidney also develops cysts.
    (E)   There are usually enough functional neurons to prevent renal insufficiency.

153

417. All of the following statements about unilateral renal agenesis are true **EXCEPT**:

(A)   condition is relatively common
(B)   males are affected most
(C)   is compatible with postnatal life
(D)   infants affected have a characteristic facial appearance
(E)   is usually not associated with oligohydramnios

During examination of the lower portion of the anterior abdominal wall of the newborn infant shown in **Figure 8.3** the clinician discovers a bright red, mucosal-like layer in the midline that separates the right and left sides of the abdominal wall. The right and left sides of the genitalia also appear separated.

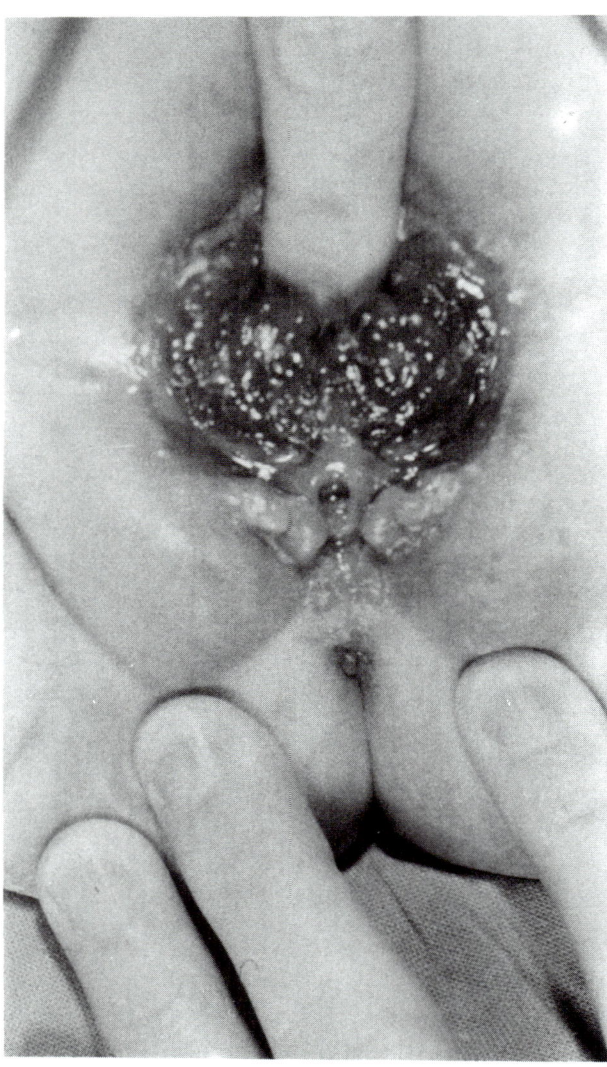

Figure 8.3

154

Choose the **BEST** response.

418. The name of this malformation is

    (A)   ruptured urachal cyst
    (B)   exstrophy of the urinary bladder
    (C)   gastroschisis
    (D)   ruptured omphalocele
    (E)   ulceration of the rectus abdominis muscles

419. All of the following statements about this kind of developmental anomaly are true **EXCEPT**:

    (A)   It occurs more often in males.
    (B)   It results when the umbilical cord ruptures.
    (C)   The red mucosal layer shown in **Figure 8.3** is the posterior wall of the urinary bladder.
    (D)   Wide separation of the pubic bones often accompanies this malformation.
    (E)   It is a relatively rare malformation.

420. This anomaly is **MOST** often associated with

    (A)   epispadias
    (B)   hypospadias
    (C)   urachal fistula
    (D)   adrenal hyperplasia
    (E)   pelvic kidney

421. Drops of a clear, yellowish fluid are observed around the umbilicus of a young infant. What is the **MOST** likely diagnosis?

    (A)   urachal cyst
    (B)   gastroschisis
    (C)   urachal fistula
    (D)   infected umbilicus
    (E)   urachal sinus

ANSWERS AND TUTORIAL ON ITEMS 416–421

The answers are: **416-C; 417-D; 418-B; 419-B; 420-A; 421-C**. In **congenital polycystic disease**, the kidney contains multiple cysts of various sizes that are dilations of nephrons. It is a relatively common disease that can cause death as a result of severe renal insufficiency. Some of the affected infants can survive with hemodialysis and kidney transplant.

Unilateral renal agenesis is relatively common, occurring about once in 1000 newborns. It occurs more often in males and it often causes no symptoms. A characteristic facial appearance and oligohydramnios are associated with bilateral absence of kidneys but not when it is absent only on one side.

The malformation shown in **Figure 8.3** is **urinary bladder exstrophy**, a relatively rare malformation (approximately 1 in 20,000 births) that occurs more often in males and is often accompanied by epispadias and wide separation of the pubic bones. The malformation is thought to be caused by incomplete formation of the lower abdominal wall in the midline that usually protects the urinary bladder. The poorly formed abdominal wall and the anterior urinary bladder wall break down thereby exposing the mucosa of the posterior urinary bladder wall to the outside. The exposed urinary bladder mucosa often appears everted and inflamed.

During early embryonic development, the **allantois** courses from the **urogenital sinus** into the umbilical cord. During the fetal period, the allantois is replaced by a thick tubular structure called the **urachus** that usually closes and transforms into a fibrous band (i.e., median umbilical ligament). Very rarely the urachus fails to close making it possible for urine to escape from the urinary bladder through the umbilicus of the newborn. Partial closure leads to the formation of **urachal cysts** and **sinuses** neither of which result in urinary leakage through the umbilicus.

<u>**Items 422-426**</u>

Choose the **BEST** response.

422.   The indifferent gonad contains

        (A)   primary sex cords
        (B)   rete testis
        (C)   tubuli recti
        (D)   tunica albuginea
        (E)   none of the above

423.   Primordial germ cells first appear in

        (A)   primary sex cords
        (B)   genital ridge
        (C)   rete testis
        (D)   secondary sex cords
        (E)   yolk sac endoderm

424.   Primary sex cords

    (A)   give rise to primordial germ cells
    (B)   are found only in the female
    (C)   breakup into rete ovarii in the female
    (D)   are derived from yolk sac
    (E)   are derived from genital ridge mesenchyme

425.   Which of the following structures is present in an embryo in the indifferent stage of genital development?

    (A)   rete testis
    (B)   primary sex cords
    (C)   cortical cords only
    (D)   paradidymis
    (E)   primordial follicles

426.   Which of the following structures undergoes an indifferent stage of development?

    (A)   external genitalia
    (B)   adrenal gland
    (C)   kidney
    (D)   urinary bladder
    (E)   ureter

ANSWERS AND TUTORIAL ON ITEMS 422-426

The answers are: **422-A; 423-E; 424-C; 425-B; 426-A**. The site of origin of the indifferent gonad is indicated by a thickening of the epithelium on the medial side of the mesonephros. Finger-like cords of epithelium called **primary sex cords** grow into the underlying mesenchyme. The rete testis, tubuli recti and tunica albuginea are structures found in the testis.

    Primordial germ cells are large, spherical cells that are visible early in the fourth week among the endodermal cells of the yolk sac. During formation of the tail fold these cells are incorporated into the embryo and by six weeks relocate to the genital ridge on each side of the hindgut where they become incorporated in the primary sex cords.

    The **primary sex cords** form in the indifferent gonad in both sexes. In the female they never become prominent but extend into the medulla of the gonad where their rudiments form the rete ovarii. Secondary sex cords or cortical cords develop subsequently as finger-like extensions from the surface epithelium and incorporate the primordial germ cells. The secondary cords form primordial follicles at about four months prenatally. In the male, because of the presence of a gene for a testis determining factor located in the Y chromosome, the primary sex cords become the rete testis in the medulla and seminiferous tubules and tubuli recti in the cortex of the gonad transforming it into a testis.

Of the structures listed, only the external genitalia undergo an indifferent stage. The internal genitalia also undergo an indifferent stage when both the mesonephric and paramesonephric ducts are present.

Items 427-431

Choose the **BEST** response

427. The uterus is formed from

   (A)   caudal part of paramesonephric ducts
   (B)   sinovaginal bulbs
   (C)   pelvic part of urogenital sinus
   (D)   caudal part of mesonephric ducts
   (E)   sinus tubercle

428. Inadequate fusion of the caudal part of the paramesonephric duct results in all of the following **EXCEPT**:

   (A)   bicornate uterus
   (B)   arcuate uterus
   (C)   septate uterus
   (D)   atresia of the uterine tube
   (E)   double uterus

429. The epithelium of the female urethra is derived from

   (A)   ectoderm
   (B)   mesoderm
   (C)   endoderm
   (D)   glandular plate
   (E)   urethral plate

158

430.   Which of the following structures is vestigial in the adult female?

    (A)   paramesonephric duct
    (B)   mesonephric tubules
    (C)   phallus
    (D)   urogenital sinus
    (E)   urogenital folds

431.   All of the following structures are derived from endoderm **EXCEPT**:

    (A)   primordial germ cells
    (B)   allantoic epithelium
    (C)   entire female urethra
    (D)   mesonephric duct
    (E)   urinary bladder epithelium

ANSWERS AND TUTORIAL ON ITEMS 427-431

The answers are: **427-A; 428-D; 429-C; 430-B; 431-D**. The **uterus** develops from fusion of the caudal parts of the **paramesonephric (Müllerian) ducts**. The cranial, unfused parts form the uterine tubes. Inadequate fusion of the caudal parts of the paramesonephric ducts results in varying degrees of separation of the uterus into two parts. Atresia of the uterine tubes results from faulty development of the cranial, unfused parts of the paramesonephric ducts.

    The **epithelium of the entire female urethra** is derived from endoderm of the **urogenital sinus**. The **mesonephric tubules** in the female form two nonfunctional structures called the **epoophoron** and the **paroophoron** that are present sometimes in the mesovarium. All of the structures listed in Item 431 are derived from endoderm except the mesonephric duct that develops from mesoderm in the lateral part of the urogenital ridge.

Items 432-439

Figure 8.4 is a view of the external genitalia of a female infant. The mother brought the infant to the pediatric clinic because of the presence of a ball-like structure on her "bottom".

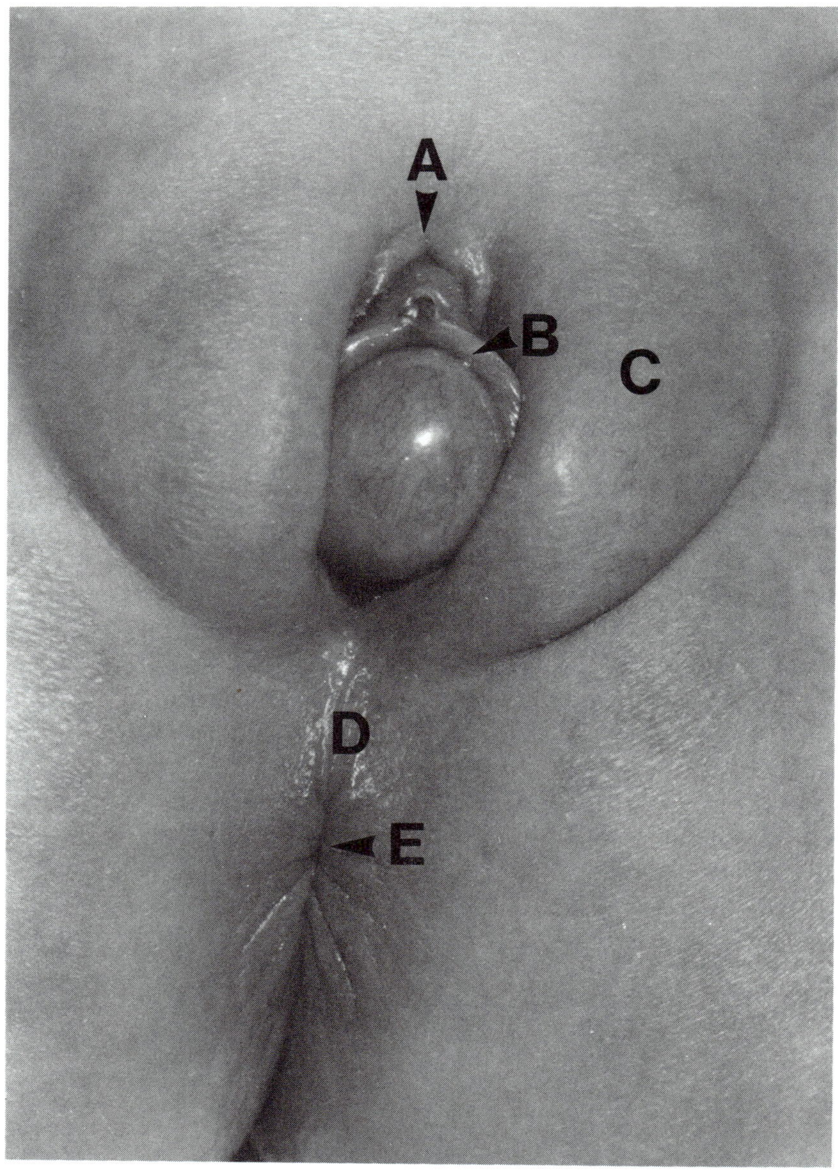

Figure 8.4

160

Choose the **BEST** response.

432. After examination of this infant, you conclude that the **MOST** likely cause of this condition is

   (A)   prolapse of the uterus
   (B)   unruptured hymen
   (C)   prolapse of the urethra
   (D)   malignant tumor
   (E)   ectopic ovary

Using the lettered areas in **Figure 8.4** choose the **BEST** response for Items 433 through 439. Answers may be used once, more than once, or not at all.

433. Develops from the urogenital fold.

434. Homologue of the scrotum in the male.

435. Forms part of the spongy urethra in the male.

436. Precursor is the genital tubercle.

437. Homologue of the glans penis.

438. Place of fusion of the urorectal septum with the cloacal membrane.

439. Precursor is the proctodeum.

ANSWERS AND TUTORIAL ON ITEMS 432-439

The answers are: **432-B; 433-B; 434-C; 435-B; 436-A; 437-A; 438-D; 439-E**. The dome-like structure in **Figure 8.4** is an unruptured **hymen** that is bulging into the perineum from the pressure of fluid that has collected in the infant's uterus and vagina. The condition is called **hydrometrocolpos**. Except for the imperforate hymen, this child's external genitalia are normal.

   The **urogenital folds** in the indifferent external genitalia give rise to the **labia minora** (B) in the female and the **spongy urethra** in the male. They remain separated in the female but in the male they fuse in the midline thereby forming the raphe on the underside of the penis. The **labioscrotal folds** are precursors of the **labia majora** (C) in the female where they remain separate. In the male they fuse in the midline to form the **scrotum**. The line of fusion of the labioscrotal folds is the **scrotal raphe**. The **genital tubercle** (i.e., phallus) becomes the **glans clitoris** (A) in the female and the **glans penis** in the male. The **perineal body** (D) or center of the perineum is the place where the urorectal septum fuses with the cloacal membrane thereby separating the anterior, urogenital area from the posterior, anorectal area. The ectodermal depression in the embryo called the **proctodeum** gives rise to the lower part of the anal canal and anus (E).

<u>Items 440-442</u>

Choose the **BEST** response.

440.　The membrane separating the lumen of the vagina from the urogenital sinus during the first trimester will develop into the

　　　(A)　vaginal vestibule
　　　(B)　vaginal fornix
　　　(C)　sinovaginal bulb
　　　(D)　hymen
　　　(E)　female urethra

441.　The vestibule of the vagina is a derivative of the

　　　(A)　urachus
　　　(B)　urogenital sinus
　　　(C)　paramesonephric ducts
　　　(D)　vagina
　　　(E)　urethra

442.　A vestigial remnant of the mesonephric duct in the female that may form cysts in the lateral wall of the uterus and vagina is

　　　(A)　Gartner's duct
　　　(B)　epoophoron
　　　(C)　paroophoron
　　　(D)　appendix vesiculosa
　　　(E)　rete ovarii

The answers are: **440-D; 441-B; 442-A**. During the first trimester the **hymen** separates the lumen of the vagina from the lumen of the **urogenital sinus**. It is formed by invagination of the posterior wall of the urogenital sinus as the lower end of the vagina expands. The lowest part of the urogenital sinus becomes the vestibule of the **vagina**.

Mesonephric (Wolffian) duct remnants sometimes persist along the lateral wall of the uterus or vagina and are referred to as **Gartner's duct**. They can also fill with fluid giving rise to cysts.

The **epoophoron, paroophoron** and **appendix vesiculosa** are also nonfunctional remnants of the mesonephric duct or tubules and are located near the uterine tubes more cranially that Gartner's duct and cysts. The **rete ovarii** is a vestige of the primary sex cords located in the medulla of the ovary.

Items 443 and 444

Choose the **BEST** response.

443. All of the following statements about female pseudohermaphroditism are true **EXCEPT**:

    (A)   usually caused by adrenal hyperplasia during the fetal period
    (B)   external genitalia are ambiguous
    (C)   androgenic agents taken during pregnancy may cause this condition
    (D)   ovaries are abnormal
    (E)   chromosome constitution is 46, XX

444. Which statement concerning individuals with androgen insensitivity syndrome (testicular feminization) is **TRUE**?

    (A)   have a 46, XX chromosome constitution
    (B)   usually have an ambiguous external genitalia
    (C)   do not develop breasts at puberty
    (D)   are raised usually as males
    (E)   have testes

The answers are: **443-D; 444-E**. Individuals with **female pseudohermaphroditism** usually have normal ovaries and a 46, XX chromosome constitution. **Congenital adrenal hyperplasia** in the fetus or the consumption of androgenic agents by the mother during pregnancy causes the ambiguous genitalia (i.e., clitoral enlargement and labial fusion). **Androgen insensitivity (testicular feminization) syndrome** occurs in genetically male individuals (i.e., have a 46, XY chromosome constitution and testes) but their external genitalia are female and they have normal development of breasts at puberty. They are raised as females but the vagina usually ends blindly and the uterus is either absent or rudimentary. The cause of the condition is thought to be a constitutive **defect in androgen receptors**.

Items 445-453

Choose the **BEST** response.

445.   The only functional derivatives of the mesonephric tubules in the male are

       (A)   efferent ductules
       (B)   rete testis
       (C)   seminiferous tubules
       (D)   paradidymis
       (E)   appendix of testis

During physical examination of a male infant, the testes could not be palpated on either side in the scrotum.

446.   Failure of descent of the testis is called

       (A)   hypospadias
       (B)   hydrocele
       (C)   cryptorchism
       (D)   epispadias
       (E)   orchitis

447. All of the following statements about undescended testis are true **EXCEPT**:

(A)   About 30% of premature male infants have the condition.
(B)   About 3% of full-term male infants have the condition.
(C)   Corrective surgery is usually required when the testis is not in the scrotum in a full-term male infant.
(D)   An undescended testis fails to mature.
(E)   An undescended testis has a considerably higher chance of becoming malignant.

448. Cells contained within the primary sex cords in the male are

(A)   primordial germ cells
(B)   cells destined to form interstitial cells
(C)   cells destined to produce testosterone
(D)   precursors of secondary sex cords
(E)   precursors of efferent ductules

449. A vestigial remnant of the mesonephric duct in the male is

(A)   appendix of epididymis
(B)   vas deferens
(C)   seminal vesicle
(D)   appendix of testis
(E)   duct of epididymis

450. Which of the following structures is derived from intermediate mesoderm?

(A)   mesonephric duct
(B)   adrenal cortex
(C)   urethra
(D)   urinary bladder
(E)   primordial germ cells

451. The prostate develops as an evagination of the

(A)   mesonephric duct
(B)   urogenital sinus
(C)   urorectal septum
(D)   ureteric bud
(E)   allantois

452.   The seminal vesicle is an outgrowth of the

    (A)   mesonephric duct
    (B)   paramesonephric duct
    (C)   Gartner's duct
    (D)   ureteric bud
    (E)   urogenital sinus

453.   The bulbourethral gland develops as an outgrowth from the

    (A)   membranous urethra
    (B)   spongy urethra
    (C)   ejaculatory ducts
    (D)   prostate gland
    (E)   vas deferens

ANSWERS AND TUTORIAL ON ITEMS 445-453

The answers are: **445-A; 446-C; 447-C; 448-A; 449-A; 450-A; 451-B; 452-A; 453-B**. The **rete testis** that develops from the primary sex cords becomes continuous with 15-20 **mesonephric tubules** that transform into **efferent ductules**. The ductules connect with the duct of the **epididymis** which is a derivative of the **mesonephric (Wolffian) duct**.

Failure of descent of the testis is called **cryptorchism**. Each testis usually passes through the inguinal canal during the seventh month of fetal life and enters the scrotum during the eighth month. Over 97% of full-term newborn males have both testes in the scrotum. Most undescended testes in such individuals spontaneously descend into the scrotum during the first three months after birth so that corrective surgery is not usually required. About 30% of premature male infants have cryptorchism. An undescended testis fails to mature resulting in **sterility**. Undescended testes also have a fifty-fold increased chance of becoming **malignant**. Therefore, palpating the presence of bilateral testes in the scrotum should be part of every pediatric physical examination.

During the sixth week the primary sex cords incorporate the primordial germ cells. Under the influence of the **testis-determining factor (TDF)** the primary sex cords become seminiferous tubules and the primordial germ cells, located in the wall of the tubules, become **spermatogonia**. The mesenchyme between the seminiferous tubules gives rise to the interstitial **Leydig cells** that produce the male sex hormone, **testosterone**.

The **appendix of the epididymis** is the only vestigial remnant of the **mesonephric duct**. The **ductus deferens**, **seminal vesicles**, and **duct of the epididymis** are functional derivatives of the mesonephric duct. The **appendix of the testis** and **prostatic utricle** are vestigial remnants of the **paramesonephric duct**.

Of the structures listed in Item 450, only the mesonephric duct is derived from **intermediate mesoderm**. The pronephric duct that runs caudally and opens into the cloaca early in the fourth week becomes the mesonephric duct late in the fourth week when tubules form in the thoracic part of the intermediate mesoderm and drain into the duct. At that time the duct opens into the part of the cloaca destined to become the urogenital sinus.

The epithelial, secretory components of the **prostate gland** are formed from multiple outgrowths of the **endodermal lining of the prostatic urethra**. This part of the male urethra is derived from the pelvic part of the **urogenital sinus**. The seminal vesicle forms as a lateral outgrowth of the caudal end of the mesonephric duct. Its mesodermally derived epithelial lining produces secretions for seminal fluid that nourishes sperm. The pea-sized **bulbourethral gland** develops as an outgrowth from the spongy part of the urethra. The gland is derived from the endodermal lining of the phallic part of the urogenital sinus and produces lubricants during sexual arousal that facilitate intromission.

Figure 8.5 is a view of the inferior aspect of the penis and scrotum of a two-month-old male infant. The mother brought the infant to the pediatric clinic because, during micturition, urine ran from the opening at the bottom of the midline groove shown in the figure instead of from the tip of the penis.

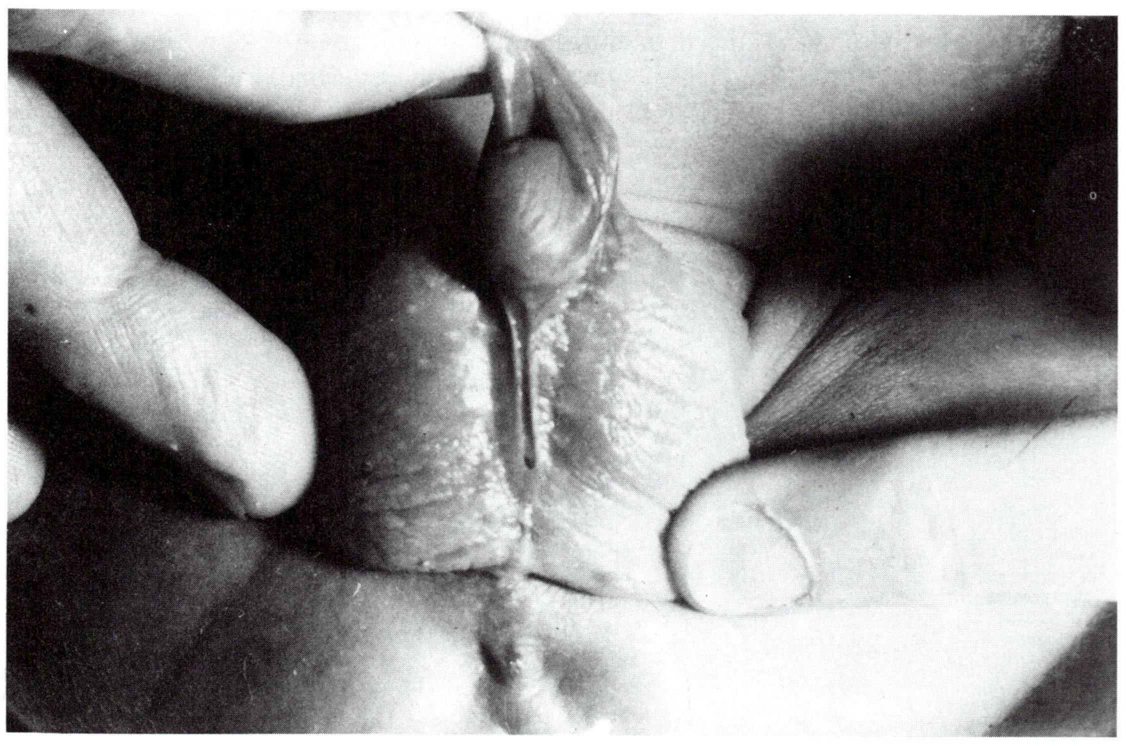

Figure 8.5

Choose the **BEST** response.

454. This condition is called

    (A)   epispadias
    (B)   cleft scrotum
    (C)   cryptorchism
    (D)   hypospadias
    (E)   micropenis

455. This condition results from failure of fusion of the

(A)   labioscrotal swellings
(B)   genital tubercle
(C)   urogenital folds
(D)   paramesonephric ducts
(E)   none of the above

456. Which of the following statements is **TRUE**?

(A)   Only one X chromosome is needed to bring about complete ovarian development.
(B)   Loss of an X chromosome interferes with the movement of primordial germ cells to the gonadal ridges.
(C)   In embryos with XXX or XXY sex chromosome constitution, the number of chromosomes is important in sex determination.
(D)   The embryo develops as a male if a normal Y chromosome is present.
(E)   All of the above are true.

457. When both testes fail to develop, all of the following events occur **EXCEPT**:

(A)   Müllerian-inhibiting factor (MIF) is absent.
(B)   Mesonephric ducts regress.
(C)   External genitalia are male.
(D)   Paramesonephric ducts do not regress.
(E)   Masculinity is repressed.

458. All of the following statements about male pseudohermaphroditism are true **EXCEPT**:

(A)   Testes are always rudimentary.
(B)   Nuclei are chromatin negative.
(C)   Chromosome constitution is 46, XY.
(D)   Intersex condition of the external genitalia is variable.
(E)   Caused by inadequate production of testosterone by the fetal testes.

ANSWERS AND TUTORIAL ON ITEMS 454-458

The answers are: **454-D; 455-C; 456-D; 457-C; 458-A**. The condition shown in **Figure 8.5** is **scrotal hypospadias** caused by inadequate midline fusion of the **urogenital folds** in male embryos. There is incomplete formation of the spongy urethra. Hypospadias results from inadequate production of androgens by the testes or inadequate receptors for the hormones. It is a frequently occurring defect (1/300 male infants). The external urethral orifice may be located anywhere on the ventral midline aspect of the glans of the penis, shaft of the penis, scrotum or perineum.

The Y chromosome regulates the production of **testis-determining factor (TDF)** that causes the medulla of the indifferent gonad to differentiate into a testis. Two X chromosomes are needed to bring about complete development of the ovary but the presence of only one X chromosome (i.e., Turner's syndrome with 45, X chromosome constitution) does not interfere with the movement of primordial germ cells to the gonadal ridge. In embryos with a XXX sex chromosome constitution, the number of X chromosomes seems to be unimportant in sex determination. The absence of a Y chromosome results in the formation of an ovary. The type of gonads determines the type of differentiation that occurs in the genital ducts and external genitalia.

When the testes fail to develop the external genitalia are ambiguous (i.e., in an intersex state). All of the other events listed occur. In **male pseudohermaphroditism**, testicular development ranges from normal to rudimentary.

Items 459 and 460

Choose the **BEST** response.

459.   Congenital adrenal (suprarenal) hyperplasia (CAH)

   (A)   Usually causes female pseudohermaphroditism.
   (B)   Results from excessive epinephrine production in the adrenal medulla.
   (C)   Is an autosomal dominant disorder.
   (D)   Leads to retarded growth.
   (E)   Does not affect males.

460.   All of the following statements about the development of the adrenal gland are true **EXCEPT**:

   (A)   The cells that form the fetal cortex are derived from mesoderm.
   (B)   The medulla differentiates from neural crest cells.
   (C)   The permanent cortex is derived from the mesothelium.
   (D)   The gland in the fetus is about the same size as the adult gland relative to body weight.
   (E)   During infancy the gland becomes smaller because the fetal cortex regresses.

170

The answers are: **459-A; 460-D**. The most common cause of female pseudohermaphroditism is **congenital adrenal hyperplasia** (CAH). This condition results from exposure of the female fetus to the excessive androgens produced by the adrenal cortex. The effects in female fetuses are mainly clitoral enlargement and labial fusion, giving the external genitalia an intersex appearance. Affected males have a normal external genitalia. The excess production of androgens leads to accelerated growth in both sexes. CAH is an autosomal recessive genetic disorder.

The **adrenal glands** in the fetus are 10 to 20 times larger they are in the adult, relative to body weight. The large size in the fetus is due to an excessively large cortex. After birth the fetal cortex regresses rapidly.

Items 461-465

Examine the photomicrograph of a placental villus in **Figure 9.1** below and then match the **MOST** appropriate description of the functional role of the labeled structures with the correct answer. Answers may be used once, more than once, or not at all.

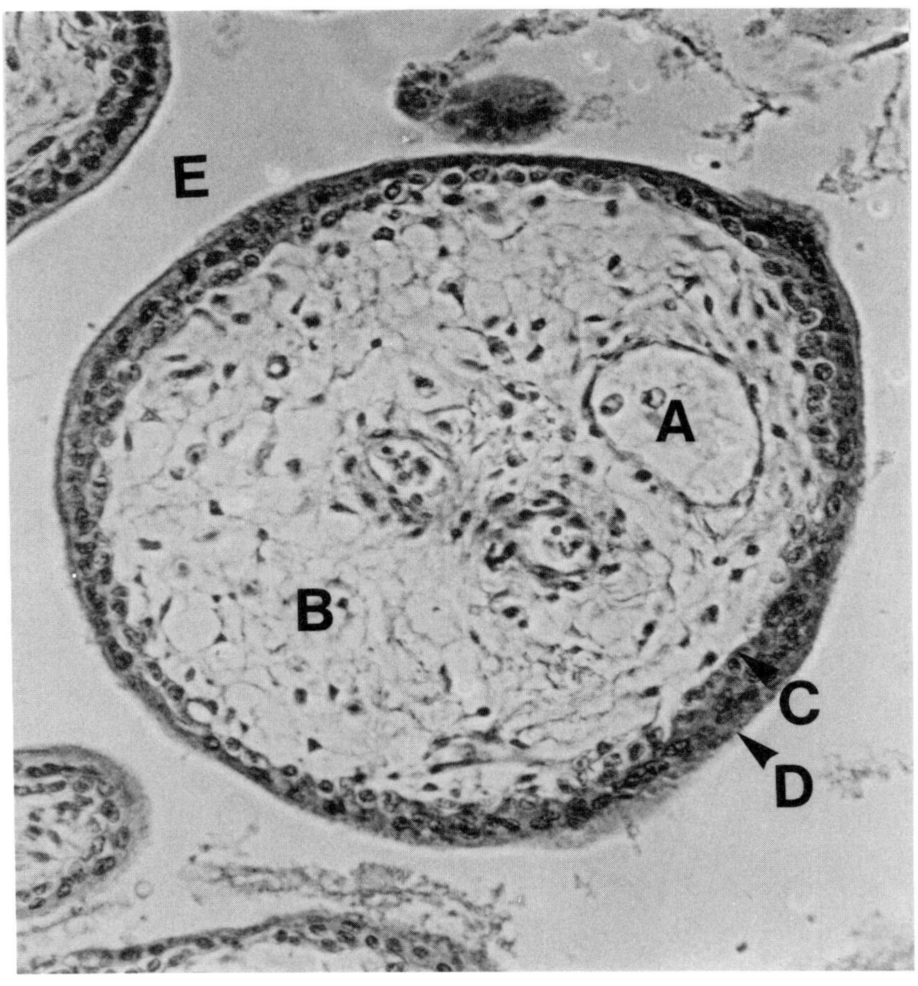

Figure 9.1

461.   This structure is formed entirely from fetal cells. It carries fetal RBCs.

462.   This structure secretes human chorionic gonadotrophin.

463.   This structure is a highly proliferative cell layer that produces the syncytiotrophoblast.

464.   This structure contains maternal blood cells.

465.   This structure is a connective tissue domain derived entirely from fetal cells.

ANSWERS AND TUTORIAL ON ITEMS 461-465

The answers are: **461-A; 462-D; 463-C; 464-E; 465-B**. **Figure 9.1** is a histological cross section of a tertiary villus from the human placenta. **Fetal blood vessels** (A) carry deoxygenated blood from the fetus to the villi where gas exchange occurs. Oxygenated maternal blood enters the **intervillous space** (E) and transfers its oxygen to fetal RBCs in fetal blood vessels. These blood vessels are supported by a core of fetally derived **connective tissue** (B). A proliferative layer of **cytotrophoblasts** (C) surrounds the connective tissue core of the villus. Daughter cells from cytotrophoblastic cell mitoses fuse with one another to form the **syncytiotrophoblast** (D), a syncytial layer at the superficial boundary of the villus. The syncytiotrophoblast is the source of **human chorionic gonadotrophin** and other placental hormones such as **human placental lactogen** and **progesterone**.

Examine the labeled diagram of a term placenta in **Figure 9.2** below and then match the labeled structures with the **MOST** appropriate description of its functional role in the items below. Answers may be used once, more than once, or not at all.

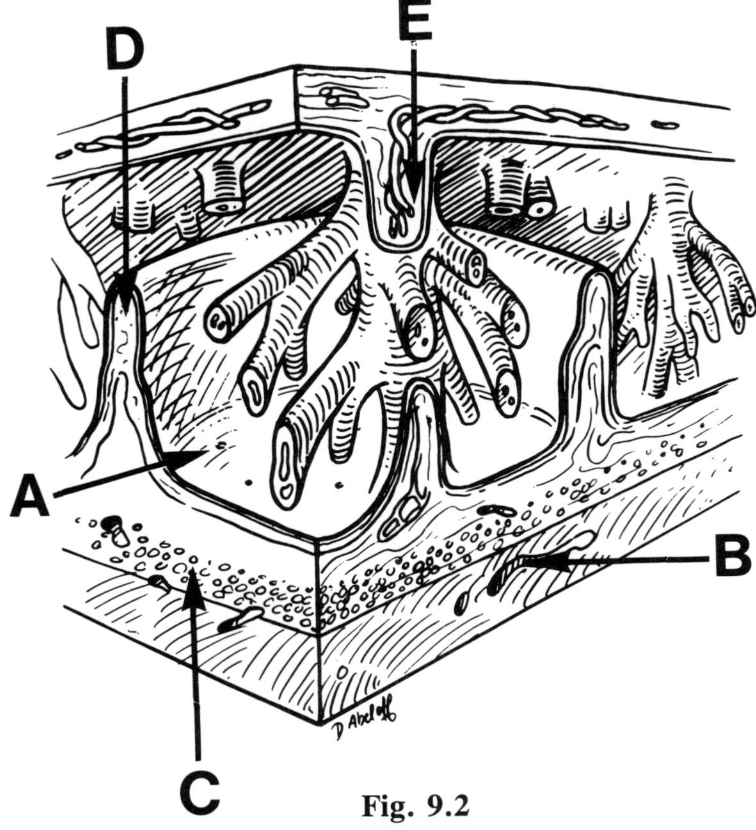

Fig. 9.2

466.   This structure joins the chorionic and decidual plates of the placenta and has branch villi on it.

467.   This structure conveys oxygenated blood to the intervillous space.

468.   This structure is a baffle which deflects blood from the intervillous space into vessels draining the intervillous space.

469.   This structure has a lumen lined with columnar endothelial cells and injects blood into the intervillous space.

470.   This structure is filled with maternal blood and branch villi.

The answers are: **466-E; 467-B; 468-D; 469-B; 470-A. Figure 9.2** is a diagram of a term human placenta. **Anchoring villi** (E) connect the chorionic plate on the fetal side of the placenta to the **decidual plate** (C) on the maternal side of the placenta. Anchoring villi have branch villi that project away from them into the **intervillous space** (A). This space receives maternal blood from the **spiral arteries** (B), which are branches of the uterine arteries. The spiral arteries have columnar endothelial cells at the lumen just before they enter the intervillous space after penetrating the decidual plate. These cells cause a sudden narrowing in the diameter of the lumen. Thus, maternal blood is injected into the intervillous space at a relatively higher pressure than that found immediately upstream in the uterine arteries. Blood streams past the villi and then is deflected by **decidual septa** (D) toward the maternal veins draining the intervillous space. Oxygen-poor fetal blood enters the placenta by way of two umbilical arteries and oxygen-rich blood drains from the placenta by way of a single umbilical vein.

Items 471-473

Choose the **BEST** response.

471.   All of the following statements are true for syncytiotrophoblast **EXCEPT**:

      (A)   It has an abundance of rough endoplasmic reticulum.
      (B)   It has little Golgi apparatus.
      (C)   It has many mitochondria.
      (D)   It secretes progesterone.
      (E)   It has many cytoplasmic transport vesicles and vacuoles.

472.   All of the following are true for cytotrophoblast **EXCEPT**:

      (A)   highly proliferative
      (B)   source of syncytiotrophoblast
      (C)   not syncytial
      (D)   decreases in amount at term
      (E)   source of human chorionic gonadotrophin

473. All of the following statements are true for the decidua **EXCEPT**:

(A)   Most decidual tissue remains behind in the uterus and is resorbed after birth.
(B)   The decidua basalis becomes part of the definitive placenta.
(C)   The decidua limits syncytiotrophoblast invasion.
(D)   The decidua parietalis forms part of the wall of the uterine cavity.
(E)   The decidua vera forms from fusion of the decidua capsularis and the decidua parietalis.

ANSWERS AND TUTORIAL ON ITEMS 471-473

The answers are: **471-B; 472-E; 473-A**. The **cytotrophoblast** is a highly proliferative layer of cells. Daughter cells from mitotic divisions fuse to form the superficial syncytiotrophoblast covering villi. As the placenta grows and the amount of syncytiotrophoblast increases, there is a concomitant decrease in amount of cytotrophoblast. At term, there are relatively few cytotrophoblastic cells remaining in the placenta.

The **syncytiotrophoblast** has a complex ultrastructure and many different cellular functions are observed. For example, the syncytiotrophoblast synthesizes and secretes a glycoprotein hormone (hCG) and thus has an abundance of rough endoplasmic reticulum (for protein synthesis) and Golgi apparatus (for protein glycosylation). There are also many mitochondria to provide ATP for protein synthesis. The mitochondria also produce ATP for the active transport of solutes and the movement of maternal IgG by membrane bound vesicles across the placenta. There are strong intercellular tight junctions and desmosomes between syncytial cells.

Decidual tissue is derived from endometrial stromal cells. Decidual cells are large, polygonal, acidophilic cells with an abundance of cytoplasmic glycogen. In some poorly understood fashion, decidual tissue limits the invasiveness of the syncytiotrophoblast. As its name implies, the decidual tissue is shed along with the placenta at birth. The **decidua basalis** forms deep to the implantation site and becomes part of the definitive placenta. The **decidua parietalis** lines the uterine cavity. The **decidua capsularis** surrounds the implantation site adjacent to the uterine lumen. As the conceptus grows, it obliterates the uterine lumen and the decidua capsularis fuses with the decidua parietalis to form the **decidua vera**.

A 42-year-old pregnant woman has chorionic villus sampling performed for cytogenetic analysis of her fetus. The scanning electron micrograph in **Figure 9.3** below is prepared from the sample. Examine the micrograph and then answer the items below concerning the developing placenta. Choose the **BEST** response.

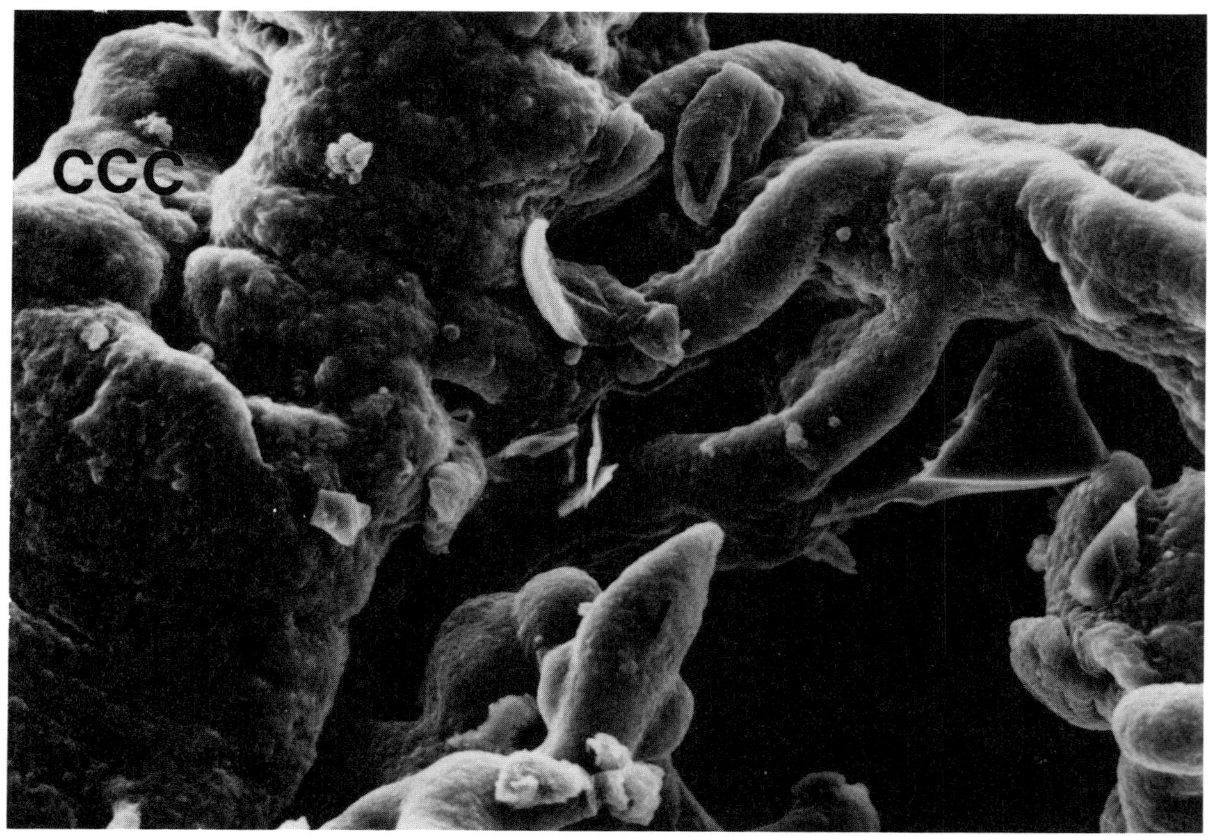

Figure 9.3

474.   The cytotrophoblastic cell columns (CCC) have all of the following morphological characteristics **EXCEPT**:

   (A)   fetal blood vessels
   (B)   maternal blood vessels
   (C)   connective tissue
   (D)   cytotrophoblast
   (E)   syncytiotrophoblast

475.   The villi (V) have all of the following morphological characteristics **EXCEPT**:

    (A)   continuous layer of cytotrophoblastic cells
    (B)   continuous syncytiotrophoblast
    (C)   connective tissue core
    (D)   decidual cells
    (E)   fetal blood vessels

476.   The space (S) normally contains

    (A)   many maternal and many fetal RBCs
    (B)   many maternal and few fetal RBCs
    (C)   no fetal RBCs
    (D)   no maternal RBCs
    (E)   many fetal and few maternal RBCs

ANSWERS AND TUTORIAL ON ITEMS 474-476

The answers are: **474-B; 475-D; 476-C**. **Figure 9.3** is a scanning electron micrograph from **chorionic villus sampling**. This technique is gradually replacing amniocentesis as a technique for gathering fetal tissue for cytogenetic and biochemical analyses. This technique has several advantages over amniocentesis:

1)   It can be performed earlier than amniocentesis and thus can be used for diagnosis at an earlier stage in pregnancy.

2)   It yields samples of rapidly growing fetal tissue which can be grown into a substantial population of cells for cytogenetic and biochemical analyses.

3)   Contamination of sample by maternal tissue is less of a problem than in amniocentesis.

Cytotrophoblastic cell columns (CCC) connect the chorionic plate and decidual (basal) plate of the developing placenta. They consist of columns of fetal connective tissue surrounding fetal blood vessels. The cytotrophoblastic cell columns contain no maternal blood vessels. They are covered by a layer of cytotrophoblast cells deep to a layer of syncytiotrophoblast at the fetal-maternal interface. **Chorionic villi** (V) project from the cytotrophoblastic cell columns into the **intervillous space** (S) which is filled with maternal RBCs. Villi are bathed in maternal blood that enters the intervillous spaces through branches of the uterine artery that penetrate the decidual plate. Villi contain fetal blood vessels surrounded by fetal connective tissue. The villi are coated by a layer of (deep) cytotrophoblast and (superficial) syncytiotrophoblast. Maternal decidual tissue is found in the placenta but not in the villi proper.

Examine **Figure 9.4** below and then choose the **BEST** response to the items below. This specimen was delivered with an apparently normal fetus. Karyotype analysis of the fetus was normal.

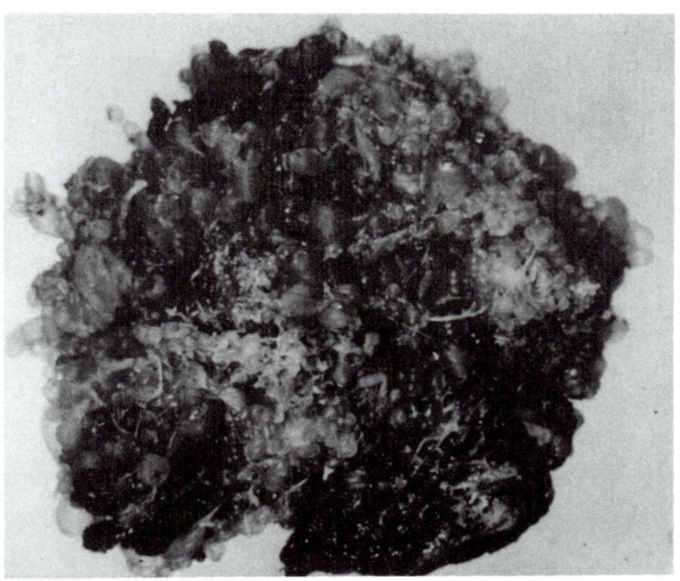

Fig. 9.4

477. This specimen is an example of

    (A)   choriocarcinoma
    (B)   normal placenta
    (C)   complete hydatidiform mole
    (D)   calcified placenta
    (E)   incomplete hydatidiform mole

478. In cases such as this cytogenetic analysis would **MOST** often reveal which of the following karyotype in the specimen?

(A)   46, XX, both X chromosomes are paternal
(B)   XY, normal male
(C)   XX, normal female
(D)   69, YYY
(E)   69, XXY

479. This condition is **MOST** prevalent in which ethnic group?

(A)   Chinese
(B)   Black-African
(C)   Northern European
(D)   Arabic
(E)   Native-American

ANSWERS AND TUTORIAL ON ITEMS 477-479

The answers are: **477-E; 478-E; 479-A**. **Figure 9.4** is an example of a **incomplete hydatidiform mole**. For all gestational trophoblastic neoplasias (GTN), the incidence is about 1 in 1500 pregnancies for Caucasians. The prevalence of all GTN and hydatidiform mole in particular is greatest in Chinese women. With incomplete molar pregnancies, fetuses are present and can be entirely normal. In these cases, the placenta usually has a triploid 69, XXY karyotype. In cases of **complete hydatidiform mole**, there is no fetus present. The karyotype is usually 46, XX with both sets of chromosomes being paternal in origin. It is thought that complete hydatidiform mole arises when an enucleate egg is fertilized by a normal X-bearing sperm. Then there is a duplication of the paternal genome, yielding a 46, XX karyotype where all chromosomes are paternally derived.

Choose the **BEST** response.

480.	Which of the following statements is **MOST** accurate for dizygotic twins?

	(A)	are usually the same sex
	(B)	are genetically identical
	(C)	can be phenotypically identical
	(D)	can share one placenta
	(E)	usually share a single amniotic sac

481.	Which of the following statements is **MOST** accurate for monozygotic twins?

	(A)	rarely share the same placenta
	(B)	can be monochorionic and monoamnionic
	(C)	can be of different sexes
	(D)	are usually genetically distinct
	(E)	are usually phenotypically distinct

482.	What percentage of naturally occurring twins are dizygotic?

	(A)	10%
	(B)	30%
	(C)	50%
	(D)	70%
	(E)	> 90%

ANSWERS AND TUTORIAL ON ITEMS 480-482

The answers are: **480-D; 481-B; 482-D. Dizygotic twins** (about 70% of the twins) are derived from two separate ova fertilized by two different sperm. Thus, dizygotic twins are no more genetically related than any other pair of siblings. They can share placentas but always have two amnions and two chorions although the chorions may be fused.

Monozygotic twins (about 30% of the twins) are derived from the division of a single zygote or cleaving zygote into two separate embryos. They are derived from a single ovum and sperm and thus are usually genetically identical. In most instances, they are essentially indistinguishable sharing finger prints, histocompatibility antigens, and personality traits. They can be diamnionic and dichorionic, diamnionic and monochorionic, or monoamnionic and monochorionic. Conjoined twins result from fusion or from incomplete division of the early embryo. Phenotypic differences can be seen in monozygotic twins, usually due to environmental factors during gestation (e.g., differences in blood supply or position in the uterus).

Examine the diagram of the ultrastructure of the syncytiotrophoblast in **Figure 9.5** below and then choose the **MOST** appropriate description of the functional role of this organelle in the term placenta. Answers may be used once, more than once, or not at all.

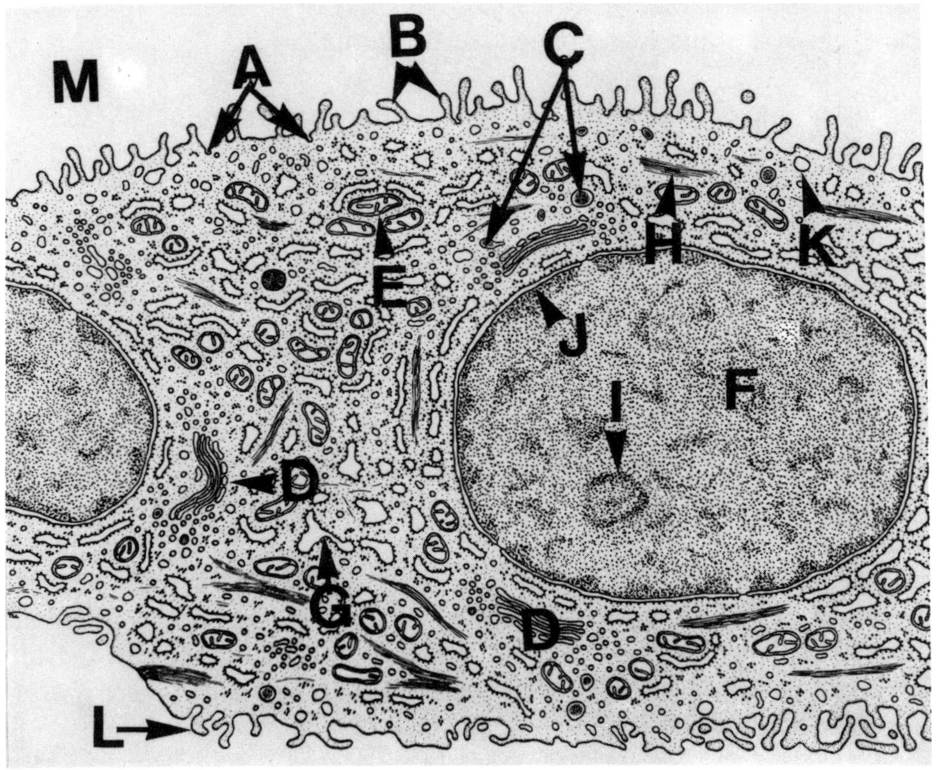

Fig. 9.5

483. This organelle is the site of glycosylation of the glycoprotein hormone human chorionic gonadotrophin.

484. This organelle is the site of synthesis for polypeptides of human chorionic gonadotrophin.

485. This organelle has shelf-like and tubular internal membranes. It is the site of some stages of steroid biosynthesis.

486.   This organelle probably replaces smooth endoplasmic reticulum for steroid biosynthesis.

487.   This organelle is involved in receptor mediated endocytosis of immunoglobulins.

488.   This organelle is the site of synthesis of macromolecules characteristically unique to the granular component of the structure labeled G.

489.   This organelle increases the absorptive surface area of the space containing maternal RBCs.

490.   This organelle increases the surface area for export of materials from the syncytiotrophoblast to the fetal connective tissue domain.

491.   This structure is closest to the basement membrane in this epithelial cell.

492.   This cytoskeletal element moves cytoplasmic organelles.

493.   Clathrin is asymmetrically distributed on this organelle.

494.   Maternal RBCs are abundant here.

ANSWERS AND TUTORIAL ON ITEMS 483-494

The answers are: **483-D; 484-G; 485-E; 486-G; 487-A; 488-I; 489-B; 490-J; 491-L; 492-H; 493-A; 494-M. Figure 9.5** is a diagram illustrating the major ultrastructural features of the syncytiotrophoblast. The superficial surface of the syncytiotrophoblast faces the **intervillous space** (L) which contains maternal RBCs. It is covered with numerous **microvilli** (B) which serve to increase the available surface area for gas and nutrient transport into the fetus and waste transport out of the fetus. **Clathrin-coated pits** (A) and their **pinocytotic vesicles** (K) are the site of receptor mediated endocytosis and transcellular transport of maternal IgGs for passive immunization of the fetus. **Cytoplasmic granules** (C) probably represent vesicles of hormones such as placental lactogen and chorionic gonadotrophin destined for secretion. The polypeptide chains of the proteins are synthesized in the **rough endoplasmic reticulum** (G) and glycosylated in the **Golgi apparatus** (D). **Mitochondria** (E) provide ATP for active transport and protein synthesis and also have some enzymes for steroid biosynthesis. The remaining enzymes for steroid biosynthesis, usually found on the smooth endoplasmic reticulum (sparse in the syncytiotrophoblast), are probably found in the rough endoplasmic reticulum. **Cytoskeletal elements** (H) move organelles about within the cell. The **nuclear envelope** (J) separates the **nucleoplasm** (F) from the cytoplasm. The **nucleolus** (I) is the site of synthesis of ribosomal RNA, a characteristic constituent of the ribosomes studding the

184

rough endoplasmic reticulum and scattered widely in the cytoplasm. On the basal surface of the syncytiotrophoblastic layer, there are numerous **surface projections** (K) that also serve to increase surface area for transport in both directions. This surface of the syncytiotrophoblast has numerous **surface projections** (L) to increase area for transport and rests on a basement membrane (not shown) in the term placenta. Earlier in pregnancy there is a continuous layer of cytotrophoblastic cells deep to the syncytiotrophoblast. The deep cytotrophoblastic cells rest on the basement membrane and the superficial syncytiotrophoblast rests on the cytotrophoblastic cells.

Items 495-496

Choose the **BEST** response

495.   All of the following statements are true concerning the umbilical cord **EXCEPT**:

        (A)   umbilical vein carries deoxygenated blood
        (B)   there are two umbilical arteries
        (C)   has mucous connective tissue rich in sulfated proteoglycans
        (D)   inserts on the chorionic plate side of the placenta
        (E)   umbilical arteries are afferent to placenta

496.   Which embryonic structure is **MOST** responsible for formation of the umbilical cord?

        (A)   extraembryonic coelom
        (B)   amnion
        (C)   chorion
        (D)   connecting stalk
        (E)   vitelline duct

ANSWERS AND TUTORIAL ON ITEMS 495 AND 496

The answers are: **495-A; 496-D**. The **umbilical cord** inserts on the **chorionic plate** of the placenta. The **basal** (decidual) **plate** of the placenta is adjacent to the uterine wall. The umbilical cord is the connection between the fetal body and the placenta, is derived from the connecting stalk and, is coated by the amnion. Two **umbilical arteries** are afferent to the placenta and there is a single efferent **umbilical vein**. Deoxygenated blood is carried from the fetus to the placenta by the umbilical arteries. Following gas exchange in the placenta, oxygenated fetal blood returns from the placenta to the fetus in the single umbilical vein.

     The length of the umbilical cord is variable but averages 55 cm at term. It can have numerous loops and is usually twisted. Compression of the umbilical vessels can compromise respiratory function for the fetus. The umbilical blood vessels are surrounded by a mucous connective tissue called **Wharton's jelly**. This peculiar connective tissue has an unusually large amount of amorphous ground substance rich in sulfated proteoglycans which keep it hydrated. The walls of the umbilical vessels are very thick due to an abundance of smooth muscle cells. These features help prevent compression of umbilical vessels.

Item 497-501

Match the placental secretion product in the list of answers below with its **MOST** appropriate functional role in the items below. Answers may be used once, more than once, or not at all.

        (A)   Chorionic gonadotrophin
        (B)   Somatomammotropin
        (C)   Progesterone
        (D)   Estriol
        (E)   Prostaglandins

497.   Structurally and functionally similar to luteinizing hormone.

498.   High levels inhibit uterine smooth muscle contraction.

499.   A growth hormone-like substance with profound effects on maternal carbohydrate utilization.

500.   Precursor is taken up by LDL-receptor on syncytiotrophoblast.

501.   Increased synthesis late in term initiates labor.

ANSWERS AND TUTORIAL ON ITEMS 497-501

The answers are: **497-A; 498-C; 499-B; 500-C; 501-E**. **Human chorionic gonadotrophin** is a glycoprotein secretion product of the syncytiotrophoblast. It is similar to luteinizing hormone and is required for maintenance of the corpus luteum of pregnancy. **Human somatomammotropin** (human placental lactogen) is similar to growth hormone and alters maternal carbohydrate metabolism to make glucose and fatty acids available to the fetus. It has little demonstrable effect on the development of the mammary glands in humans. The syncytiotrophoblast has little smooth endoplasmic reticulum but nevertheless produces and secretes **progesterone** and **estriol**, an estrogen. Progesterone is synthesized from maternal cholesterol which is taken up from maternal blood low-density lipoprotein (LDL) by a specific receptor on the syncytiotrophoblast. Progesterone is thought to inhibit contraction of uterine smooth muscle. The placenta also secretes **prostaglandins** which stimulate uterine smooth muscle contraction. The increase in prostaglandin-secreting activity of the placenta at term is probably the trigger that initiates labor.

Figure 9.6 below is a diagram of a term fetus surrounded by the fetal membranes and the uterine wall. Match the labeled structures in the diagram with the **MOST** appropriate description of this structure in the items below. NOTE: Arrowheads surrounded by a circle represent a space. Answers may be used once, more than once, or not at all.

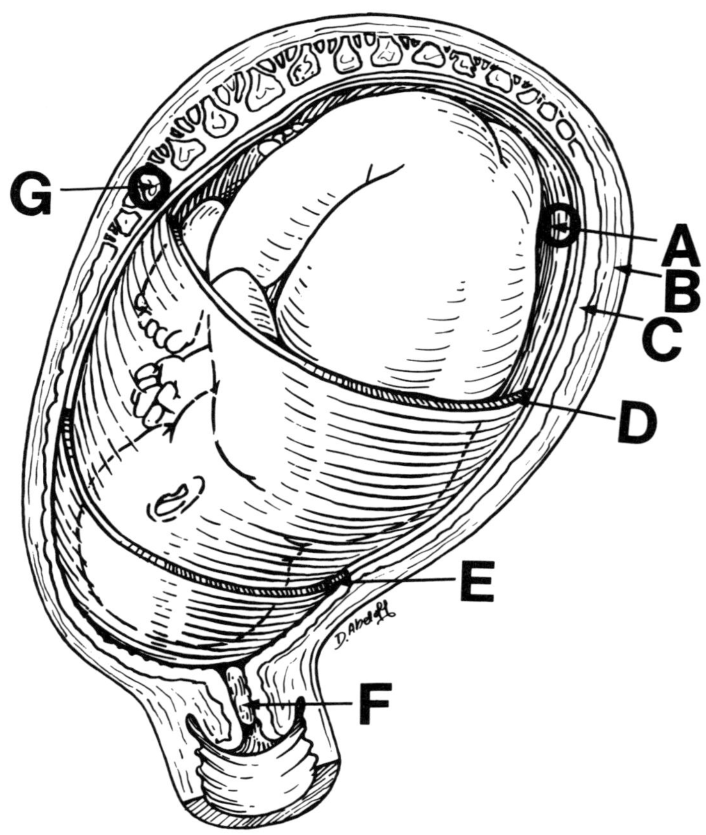

Figure 9.6

502. This structure prevents bacteria from entering the uterine cavity.

503. During amniocentesis which structure (s) would be penetrated by the needle?

   (A)   B and C only
   (B)   D and E only
   (C)   B, C, D, and E
   (D)   B, C, and D only
   (E)   C, D, and E only

504. This structure is coated on its inner aspect (facing the fluid- and fetus-filled cavity) by a cuboidal, fluid-transporting epithelium.

505. This structure is derived from the trophoblast. Its outer surface (facing the uterine wall) is covered by cytotrophoblastic cell columns and villi.

506. This space contains maternal blood.

507. This space contains amniotic fluid.

508. This decidual tissue is derived from modified endometrial stromal fibroblasts.

Choose the **BEST** response.

509. Which of the following statements **MOST** accurately describes the location or condition of the placenta shown in **Figure 9.6**?

    (A)   in the lower uterine segment
    (B)   incomplete placenta previa
    (C)   complete placenta previa
    (D)   frankly abrupted
    (E)   normal

The answers are: **502-F; 503-C; 504-D; 505-E; 506-G; 507-A; 508-C; 509-E**. **Figure 9.6** is a diagram of a term fetus contained within the uterus and surrounded by fetal membranes. The placenta is in the upper, posterior portion of the uterus, its normal location. **Placenta previa** would either partially (incomplete) or completely cover the cervical os which is filled with a **cervical mucus plug** (F) to prevent bacteria in the nonsterile vagina from entering the sterile uterine cavity. The fetus is bathed in amniotic fluid which is contained in the **amniotic cavity** (A). The **amnion** (D) borders the amniotic cavity, is coated on its inner aspect by cuboidal, fluid-transporting epithelium and is fused with the **chorion** (E).

For amniocentesis, a needle is inserted through the myometrium of the uterus (B), then the **decidual layer** (C) (formed from modified, maternal, endometrial fibroblasts), then the **chorion** (E), then the **amnion** (D) and into the **amniotic cavity** (A) where fluid containing fetal cells for karyotype analysis can be aspirated.

Oxygenated maternal blood courses from branches of the uterine arteries into the **intervillous spaces** (G) of the placenta. Here, placental villi on the surface of the chorion are bathed in maternal blood. Gas exchange occurs so that fetal blood looses CO_2 to maternal blood and takes up O_2 from maternal blood.

CHAPTER X
CAUSES OF HUMAN BIRTH DEFECTS

Items 510-514

Examine the karyotype in **Figure 10.1** and then choose the **BEST** response to the items below.

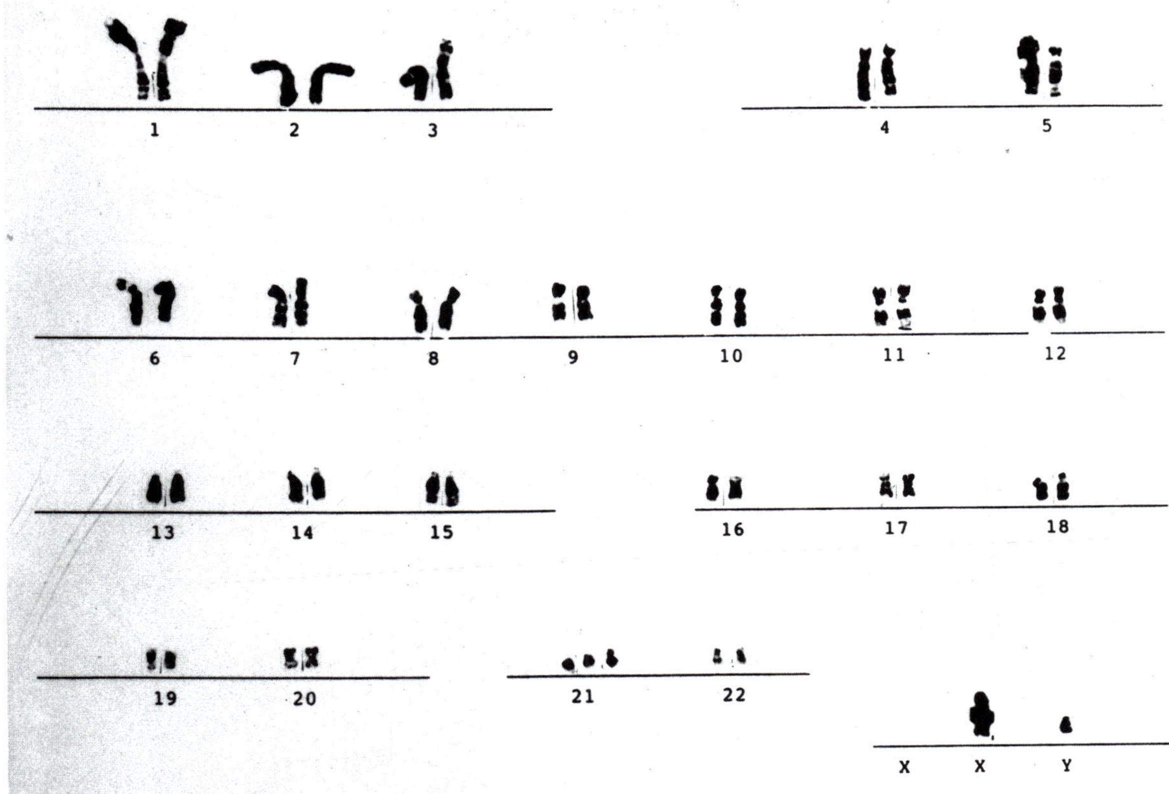

Figure 10.1

510. The developmental process whose disruption produces this karyotype is

    (A)   mitosis
    (B)   synapsis
    (C)   differentiation of primordial germ cells
    (D)   meiotic nondisjunction
    (E)   migration of primordial germ cells

511. Which shorthand notation is **MOST** appropriate for describing this karyotype?

(A) 46, XY
(B) 47, XX, +18
(C) 47, XY, +21
(D) 47, XXY
(E) 45, X

512. The **MOST** appropriate diagnosis of a child with this karyotype is

(A) Down syndrome
(B) cri du chat syndrome
(C) Marfan syndrome
(D) Klinefelter syndrome
(E) Turner syndrome

513. All of the following symptoms and characteristics are associated with this syndrome **EXCEPT**:

(A) shortened life expectancy
(B) cardiovascular anomalies
(C) short stature
(D) normal mental function
(E) oblique palpebral fissures

514. The **MOST** predictive maternal characteristic for occurrence of this birth defect is

(A) alcoholism
(B) diabetes mellitus
(C) obesity
(D) hypertension
(E) age greater than 40 years

ANSWERS AND TUTORIAL ON ITEMS 510-514

The answers are: 510-D; 511-C; 512-A; 513-D; 514-E. **Figure 10.1** is a karyotype of a male child with primary nondisjunctional trisomy 21 (**Down syndrome**; 47, XY, +21). During synapsis of the first meiotic metaphase, homologous chromosomes pair. If this pairing leads to failure of separation of chromosomes, gametes are produced with either an extra copy of some particular chromosome or the lack of that copy of the chromosome. A human female is born with all of her oocytes arrested in the first meiotic metaphase. Meiosis I is not

192

completed in the human female until just before ovulation. Consequently, synapsis can last for many years. Increasing maternal age leads to a higher incidence of nondisjunctional chromosomal anomalies including Down syndrome. When nondisjunction occurs during oogenesis, the ovum contains either 2 copies of chromosome 21 or 0 copies of chromosome 21, instead of the normal 1 copy of chromosome 21. When an ovum containing 2 copies of chromosome 21 is fertilized by a haploid sperm containing 1 copy of chromosome 21, 3 copies (2 maternal and 1 paternal) are found in the zygote and Down syndrome eventually results. The common characteristics of children with Down syndrome are short stature, straight hair, protruding tongue, oblique palpebral fissures and cardiovascular anomalies such as ventricular septal defects (VSDs). Children with Down syndrome have a shortened life expectancy and reduced mental function with an IQ typically in the 40-80 range.

Items 515-518

Examine the karyotype in **Figure 10.2** and then choose the **BEST** response to the items below.

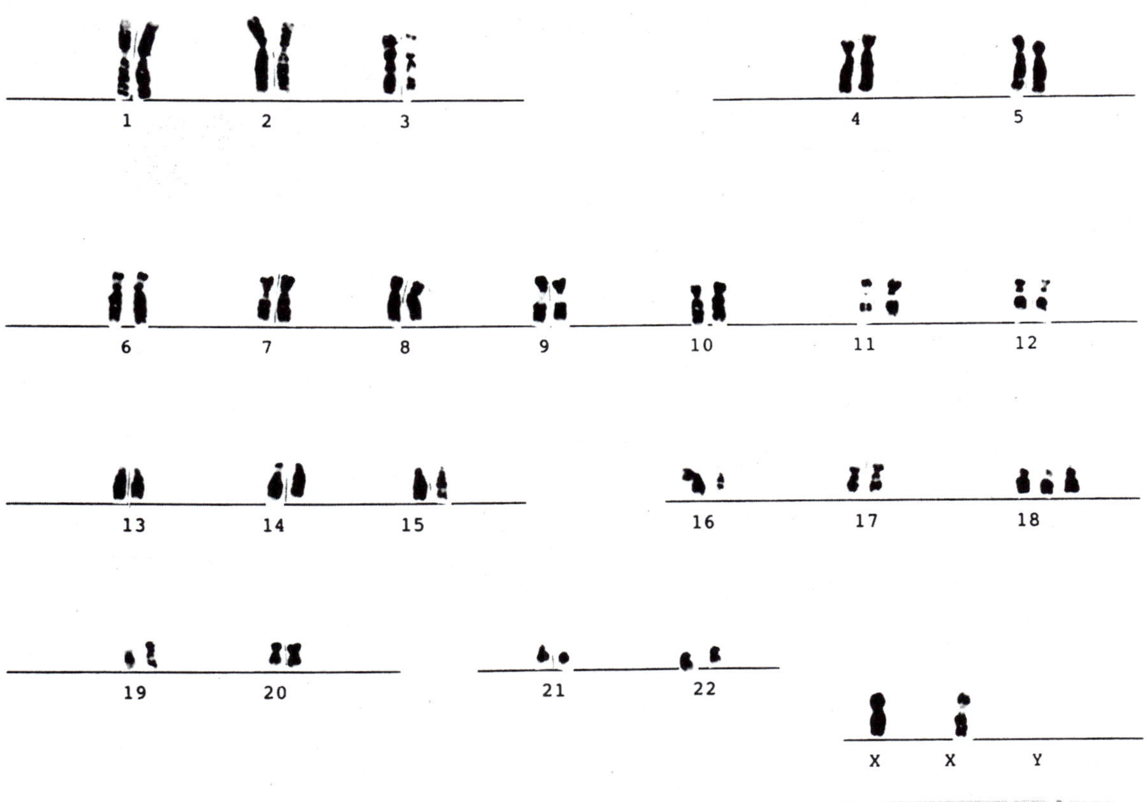

Figure 10.2

515. Which shorthand notation is **MOST** appropriate for describing this karyotype?

    (A)   46, XY
    (B)   47, XX, +18
    (C)   47, XY, +21
    (D)   47, XXY
    (E)   45, X

516. The **MOST** appropriate diagnosis of a child with this karyotype is

    (A)   Down syndrome
    (B)   Edwards syndrome
    (C)   Marfan syndrome
    (D)   Patau syndrome
    (E)   Klinefelter syndrome

517. Which of the following symptoms is **MOST** characteristic of this aneuploid disorder?

    (A)   long life span
    (B)   normal heart
    (C)   cleft palate
    (D)   normal mental function
    (E)   prominent occipital region of skull

518. The **MOST** predictive maternal characteristic for occurrence of this birth defect is

    (A)   age greater than 40 years
    (B)   cocaine abuse
    (C)   hypertension
    (D)   high fever during pregnancy
    (E)   alcoholism

ANSWERS AND TUTORIAL ON ITEMS 515-518

The answers are: **515-B; 516-B; 517-E; 518-A**. **Figure 10.2** is a karyotype of a female child with primary nondisjunctional trisomy 18 (**Edwards syndrome**). This congenital anomaly occurs in approximately 1/8,000 live-born children. The appropriate shorthand notation designating this kind of aneuploidy is 47, XX, +18. Increasing maternal age leads to a higher incidence of nondisjunctional chromosomal anomalies including Edwards syndrome. When nondisjunction occurs during oogenesis, the ovum contains either 2 copies of chromosome 18 or 0 copies of chromosome 18, instead of the normal 1 copy of chromosome

194

18. When an ovum containing 2 copies of chromosome 18 is fertilized by a haploid sperm containing 1 copy of chromosome 18, 3 copies (2 maternal and 1 paternal) are found in the zygote, resulting in trisomy 18 (Edwards syndrome). Monosomy 18 conceptuses abort spontaneously early in their development. The common symptoms peculiar to trisomy 18 are microcephaly, prominent occiput, low-set pointed ears, micrognathia, overlapping fingers, and rounded feet with a large calcaneus. Cardiovascular and renal defects are also commonly found in Edwards syndrome but are found in other conditions as well, e.g., cardiovascular defects are also common in Down syndrome. Children with trisomy 18 usually die soon after birth.

Items 519-522

Examine the karyotype in **Figure 10.3** and then choose the **BEST** response to the items below.

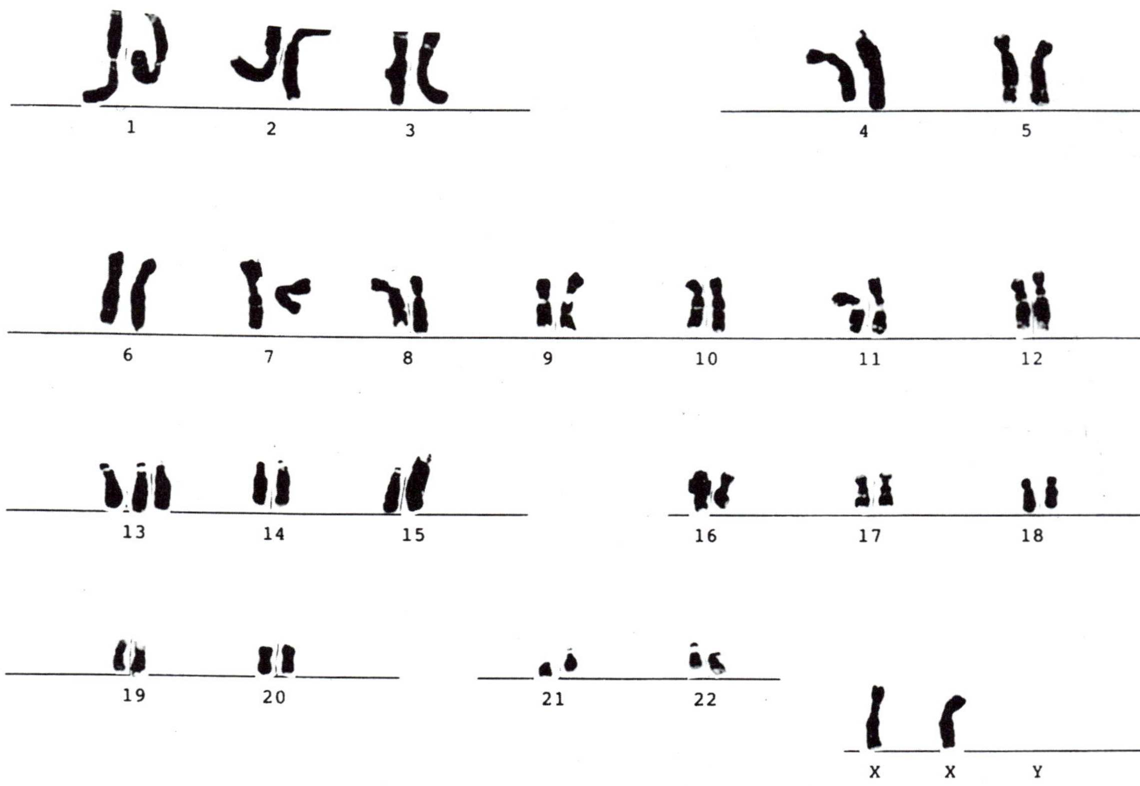

Figure 10.3

519.   Which shorthand notation is **MOST** appropriate for describing this karyotype?

   (A)   46, XX
   (B)   47, XX, +18
   (C)   47, XY, +13
   (D)   48, XXXY
   (E)   46, XY

196

520. The **MOST** appropriate diagnosis of a child with this karyotype is

    (A)   Down syndrome
    (B)   Edwards syndrome
    (C)   Patau syndrome
    (D)   fetal alcohol syndrome
    (E)   Turner syndrome

521. Which of the following symptoms is **MOST** characteristic of this aneuploid disorder?

    (A)   normal life span
    (B)   cardiovascular anomalies
    (C)   cleft palate and microphthalmia
    (D)   normal mental function
    (E)   prominent occipital region of skull

522. The **MOST** predictive maternal characteristic for occurrence of this congenital birth defect is

    (A)   heroin abuse
    (B)   age greater than 40 years
    (C)   pelvic inflammatory disease
    (D)   eclampsia
    (E)   cytomegalovirus infection

ANSWERS AND TUTORIAL ON ITEMS 519-522

The answers are: **519-C; 520-C; 521-C; 522-B**. **Figure 10.3** is a karyotype of a female child with primary nondisjunctional trisomy 13 (**Patau syndrome**). This congenital anomaly occurs in approximately 1/20,000 live-born children. The appropriate shorthand notation designating this kind of aneuploidy is 47, XX, +13. Increasing maternal age leads to a higher incidence of nondisjunctional chromosomal anomalies including Patau syndrome. When nondisjunction occurs during oogenesis, the ovum contains either 2 copies of chromosome 13 or 0 copies of chromosome 13, instead of the normal 1 copy of chromosome 13. When an ovum containing 2 copies of chromosome 13 is fertilized by a haploid sperm containing 1 copy of chromosome 13, 3 copies (2 maternal and 1 paternal) are found in the zygote resulting in trisomy 13 (Patau syndrome). Monosomy 13 conceptuses abort spontaneously early in their development. The common symptoms peculiar to trisomy 13 are holoprosencephaly, microphthalmia, anophthalmia, cleft lip, cleft palate, and polydactyl. Congenital heart defects are also commonly found in Patau syndrome but are found in other conditions as well, e.g., in Down syndrome and Edwards syndrome. Children with trisomy 13 usually die soon after birth.

Examine the karyotype in **Figure 10.4** and then choose the **BEST** response to the items below.

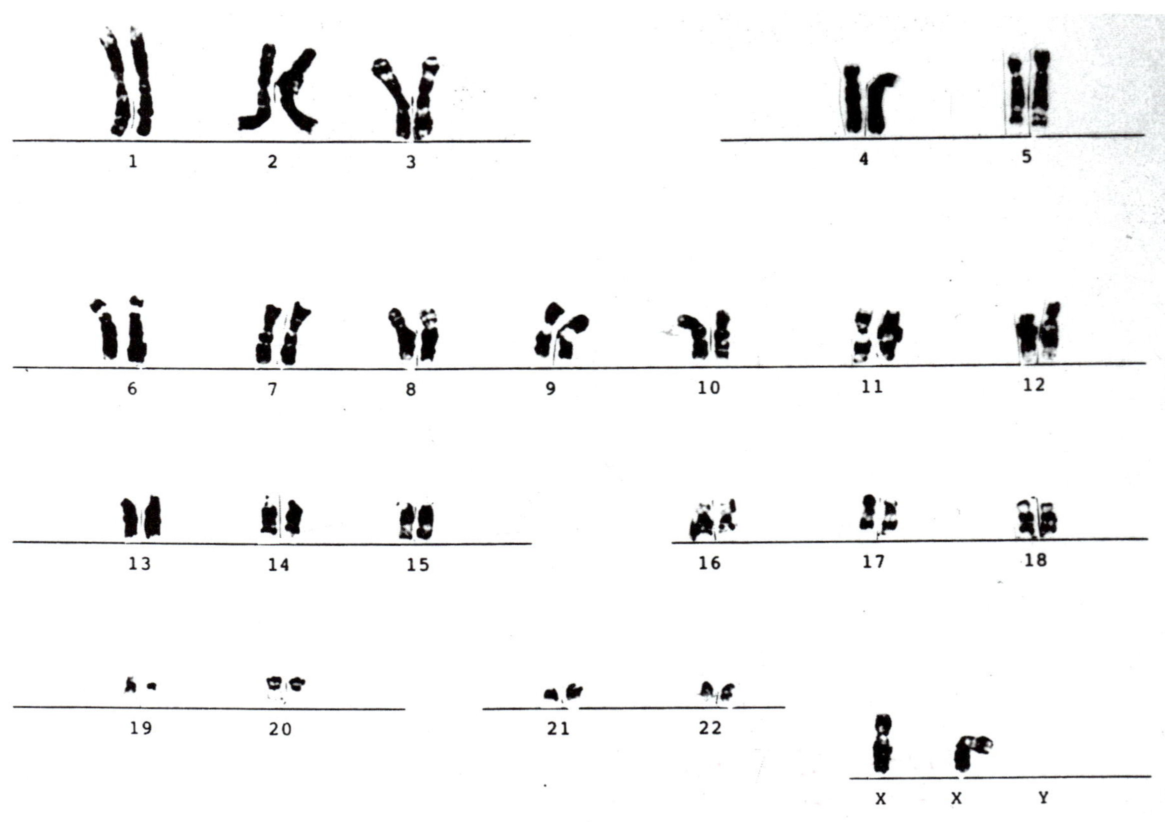

Figure 10.4

523. Which prenatal diagnostic technique is **MOST** commonly used in karyotype analysis?

(A)   ultrasound
(B)   amniocentesis
(C)   fetal heart monitoring
(D)   measurement of maternal serum α-fetoprotein
(E)   urinalysis

524. Which drug is **MOST** commonly used in karyotype analysis?

(A) steroids
(B) β-blockers
(C) colchicine
(D) ACTH
(E) ibuprofen

525. Which shorthand notation is **MOST** appropriate for describing this karyotype?

(A) 46, XX
(B) 47, XXY
(C) 46, XY
(D) 45, X
(E) 45, Y

526. Which statement **BEST** characterizes this karyotype?

(A) aneuploidy with numerical anomalies in sex chromosomes
(B) aneuploidy with numerical anomalies in autosomes
(C) euploidy
(D) monosomy
(E) trisomy

527. Which of the following disorders could be associated with this karyotype?

(A) testicular feminization syndrome
(B) Klinefelter syndrome
(C) sickle cell anemia
(D) Turner syndrome
(E) Down syndrome

The answers are: **523-B; 524-C; 525-A; 526-C; 527-C**. **Figure 10.4** is an example of a **normal female karyotype**. Fetal tissue samples are gathered for karyotype analysis. In prenatal diagnosis, fetal tissue cells can be gathered by **amniocentesis** or chorionic villus sampling. Following collection of fetal cells, they can be grown in tissue culture. Treatment of cultures with the drug **colchicine** results in the arrest of mitosis at metaphase. Chromosomes on the metaphase plate can then be spread, stained, imaged with a microscope attached to an image analyzing computer, and arranged by homologous pairs into a karyotype.

The most appropriate shorthand notation for this normal female karyotype is 46, XX. This is a **euploid** (normal number) karyotype. Aneuploid karyotypes (abnormal number) have numerical anomalies so that there are more or less than 46 chromosomes arranged as 22 pairs of autosomes and a pair of heteromorphic sex chromosomes (XX for female, XY for male). Trisomy and monosomy are examples of aneuploid karyotypes. Trisomy 21 (Down syndrome), 47, XX, +21 in a female is an example of an **autosomal aneuploidy**. Testicular feminization syndrome, a condition due to congenital absence of the testosterone receptor, is characterized by a female body habitus and gender identity with degenerate testes and a male euploid karyotype (46, XY). Turner syndrome is due to a lack of a second X chromosome (45, X) and is an example of **sex chromosome aneuploidy**, as is Klinefelter syndrome (47, XXY). Sickle cell anemia is due to a point mutation in the hemoglobin gene and would be associated with either a normal male (46, XY) or female (46, XX) karyotype.

Examine the karyotype in **Figure 10.5** and then choose the **BEST** response to the items below.

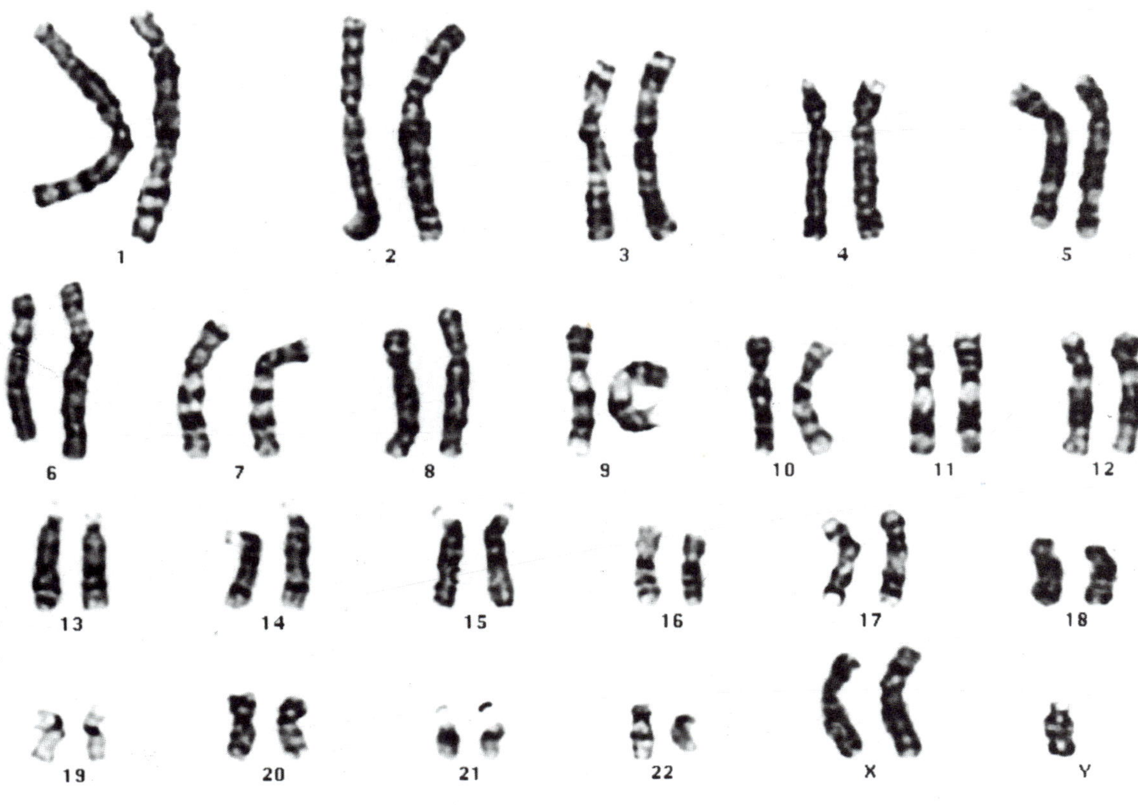

Figure 10.5

528.   Which clinical syndrome is associated with this karyotype?

    (A)   Down syndrome
    (B)   Klinefelter syndrome
    (C)   Turner syndrome
    (D)   Edwards syndrome
    (E)   Patau syndrome

529. Which phrase would **MOST** appropriately describe the testis of such a person?

(A) hypertrophied
(B) absent
(C) frequently undescended with hyalinization of seminiferous epithelium
(D) lacking Leydig cells
(E) hermaphroditic

530. All of the following characteristics are found in such individuals **EXCEPT**:

(A) profound mental retardation
(B) gynecomastia
(C) dyslexia
(D) undescended testes
(E) tall stature

531. The incidence of this condition among live-born males is approximately

(A) 1/100
(B) 1/1,000
(C) 1/5,000
(D) 1/10,000
(E) 1/100,000

ANSWERS AND TUTORIAL ON ITEMS 528-531

The answers are: **528-B; 529-C; 530-A; 531-B**. **Figure 10.5** is an example of the karyotype associated with **Klinefelter syndrome**. This was the first human sex chromosome aneuploidy that was described and is clearly the result of nondisjunction of sex chromosomes. These individuals appear physically normal at puberty. They are usually tall and thin. At puberty, signs of hypogonadism become apparent with small testes and underdeveloped secondary sexual characteristics. Histologically, the frequently undescended testes may exhibit hyalinization of the seminiferous epithelium due to prolonged exposure to core body temperature. The seminiferous epithelium requires the slightly lower temperature present in the scrotum for normal spermatogenesis. Increased incidence of seminomas, gynecomastia, and dyslexia are often associated with Klinefelter syndrome. The overall incidence is about 1/1000 male live-births.

<u>**Items 532-545**</u>

Match the **MOST** appropriate shorthand notation for karyotypes in the answers below with the description in the items below. Answers may be used once, more than once, or not at all.

    (A)   45, X
    (B)   47, XXY
    (C)   47, XYY
    (D)   47, XX, +21
    (E)   47, XX, +18
    (F)   47, XX, +13
    (G)   47, XXX
    (H)   47, XY, +16

532. Characterized by short stature, streak gonads, neck webbing, and widely spaced nipples.

533. This karyotype is more commonly found among males in prisons and mental hospitals in individuals greater than 6 feet tall.

534. May present as primary or secondary amenorrhea.

535. Characterized by severe mental retardation, rocker-bottom feet, and an incidence of cardiac anomalies approaching 90% in affected individuals.

536. Associated with low values of maternal serum α-fetoprotein (MSAFP), unconjugated estriol (uE) and human chorionic gonadotropin (hCG).

537. Associated with low values of MSAFP and uE, increased values of hCG.

538. Characterized by the presence of hypotonia, a transverse palmar crease, mental retardation, and an increased incidence of duodenal atresia, tracheoesophageal fistula, and congenital heart disease.

539. Characterized by midline defects such as cleft lip and palate, growth retardation, severe mental retardation, and CNS anomalies such as holoprosencephaly.

540. Appear physically normal until puberty when hypogonadism is noted. Gynecomastia may be present.

541. Two Barr bodies are visible in the cells of these individuals.

542. Above average in stature, phenotypically normal, frequently mentally retarded.

543. Always female and frequently diagnosed on prenatal ultrasound by the presence of a cystic hygroma.

544. Found in 1% of live-births in an unscreened population when maternal age is 40 and above.

545. Most common trisomy seen in abortuses but never seen in live-borns.

ANSWERS AND TUTORIAL ON ITEMS 532-545

The answers are: 532-A; 533-C; 534-A; 535-E; 536-E; 537-D; 538-D; 539-F; 540-B; 541-G; 542-C; 543-A; 544-D; 545-H. **Aneuploidy** refers to any chromosome number which is not a multiple of the haploid chromosome number. It is the most common and clinically significant type of chromosome disorder. The most common mechanism is **meiotic nondisjunction** which represents failure of a pair of chromosomes to disjoin in the normal way. This usually occurs in the first meiotic division. If the error occurs during meiosis I, the gamete with 24 chromosomes contains both the maternal and paternal members of the pair. On the other hand, if it occurs during meiosis II, both copies of the nondisjoined chromosome will be either maternal or paternal. Nondisjunction can also occur in a mitotic division after zygote formation. Early mitotic nondisjunction may result in mosaicism.

The overall incidence of chromosomal abnormalities is about .7% in the live-born population. The incidence of aneuploidy is considerably higher at amniocentesis and even higher earlier in gestation, at the time of chorionic villus sampling (CVS). Obviously, a significant number of aneuploid fetuses diagnosed early in gestation would be lost spontaneously as gestation advances. The highest incidence of aneuploidy is found in abortuses with an incidence of 40-50%. **Trisomy 16** (H) is the most frequently seen trisomy in abortuses but is not found in live-born fetuses.

The incidence of aneuploidy is significantly related to **maternal age**. Total risk for chromosome abnormalities is about 1/526 for a woman of age 20 at the time of delivery. At age 35 her risk is about 1/200. By age 45, a woman's risk of all chromosome abnormalities is about 1/20, increasing to 1/7 by age 49. It is apparent that increased maternal age is by far the most common indication for invasive prenatal testing.

Aneuploidy may involve either the autosomes or the sex chromosomes. By far the most common of the autosomal trisomies in live-borns is **trisomy 21 or Down syndrome** (D). The trisomy 21 phenotype consists of prominent epicanthal folds, oblique palpebral fissures, hypotonia, short stature, a flat occiput, a short neck, a flat nasal bridge, and low-set ears. A single transverse palmar crease (simian crease) may also be present. Mental retardation is present with the average IQ measuring 40-80. Congenital heart disease is common, occurring in about one third of infants with trisomy 21. Duodenal atresia and tracheoesophageal fistulas are seen with greater frequency in these infants.

204

Trisomy 13 (Patau syndrome) (F) and **trisomy 18 (Edwards syndrome)** (E) are other autosomal trisomies seen in live-borns. Trisomy 13 is almost always lethal. It is characterized by profound mental retardation. Infants with Patau syndrome frequently have midline defects such as cleft lip and palate. CNS malformations such as holoprosencephaly or arhinencephaly are seen with increased frequency. There may be hypertelorism, microphthalmia and even absence of the eyes. Rocker-bottom feet, clenched fists and simian creases are present. Cardiac and genitourinary anomalies are common.

Trisomy 18 is characterized by intrauterine growth restriction, profound mental retardation and is almost invariably lethal. Cardiac malformations are more commonly seen in trisomy 18 than any of the other trisomies. The infants are hypertonic, with a prominent occiput, mandibular hypoplasia, and low-set ears. The fists are clenched with the second and fifth digits overlapping the third and fourth. This is known as clinodactyly. The feet are rocker-bottom. The diagnosis of trisomy 18 has been suggested with the combination of low levels of MSAFP, unconjugated estriol, and hCG.

Disorders of the sex chromosomes include 45, X; 47, XXX; 47, XXY; and 47, XYY. The first human sex chromosome aneuploidy described was **Klinefelter syndrome**, or 47, XXY (B). These individuals appear physically normal until puberty. They are usually tall and thin. At puberty signs of hypogonadism become apparent with small testes and underdeveloped secondary sexual characteristics. **Gynecomastia** is also observed occasionally. The overall incidence is about 1/1000 male live-births. A high percentage of these patients are **dyslexic**.

Individuals with the 47, XYY karyotype (C) occupy a significantly greater proportion of the population of prison inmates and male inhabitants of mental hospitals. This is especially true if one examines the group over six feet tall where the incidence may be greater than twenty percent. These individuals are of normal intelligence and have no distinguishing features. It is clear that the 47, XYY karyotype results from paternal nondisjunction.

Turner syndrome or 45, X (A) occurs in about 1/5000 female live-births. About one fourth of Turner syndrome patients are mosaic. These patients are invariably of short stature, have webbing of the neck, a broad or "shield" chest, and widely spaced nipples. The gonads are dysgenetic and appear as streaks. The patients may experience primary or secondary amenorrhea. Intellect is normal. Coarctation of the aorta is significantly more common in patients with Turner syndrome than in the general population. The diagnosis may be made prenatally on ultrasound exam where a **cystic hygroma** may be noted. This is thought to result from failure of the development of a communication between the jugular lymphatic sac and the jugular vein at about 40 days post-conception. Dilation of the jugular lymphatic sac leads to formation of cystic structures in the cervical region. Cystic hygromas are also seen with other chromosome abnormalities such as trisomy 21.

Trisomy X or 47, XXX (G) females are usually taller than average but are phenotypically normal. Many are infertile. IQ is frequently low and about 70% of patients diagnosed with trisomy X have significant learning problems. As expected, almost all cases result from maternal nondisjunction. Since all but one of the X chromosomes are inactivated, two **Barr bodies** will be observed microscopically, instead of the usual one in 46, XX females.

Match the set of symptoms in the items below with the **MOST** appropriate teratogenic agent in the list of answers. Answers may be used once, more than once, or not at all.

(A) Thalidomide
(B) Alcohol
(C) Cocaine
(D) Heroin
(E) Rubella virus
(F) Herpes simplex virus
(G) *Toxoplasma gondii*
(H) Organic mercurial compounds
(I) Synthetic progestins
(J) High doses of ionizing radiation

546. Midfacial hypoplasia, stunted growth, mental retardation

547. Masculinization of female external genitalia

548. Amelia and phocomelia

549. Cataracts and deafness

550. Spontaneous abortion, often soon after acute exposure

551. Chorioretinitis, microcephaly, and calcifications around the choroid plexus

552. Probably not teratogenic, but exposure of fetus during delivery can be fatal

ANSWERS AND TUTORIAL ON ITEMS 546-552

The answers are: **546-B; 547-I; 548-A; 549-E; 550-J; 551-G; 552-F**. **Thalidomide** (A) was widely used in Germany and other countries as a tranquilizer and sedative between 1957 and 1963. It was subsequently shown to be a potent teratogen, causing an epidemic of previously rare limb defects such as phocomelia and amelia. When use of this drug was discontinued, the incidence of these limb defects once again declined to a very low level.

Alcohol (B) is a potent teratogen. Ingestion of large quantities of alcohol, especially during the early stages of development, can lead to abnormal induction of the forebrain and a severe developmental defect known as holoprosencephaly. In addition, a less severe fetal alcohol syndrome has been identified. Afflicted children exhibit growth restriction, midfacial hypoplasia, elongated philtrum and reduced upper lip, mental retardation, hyperactivity, and speech delay. Women who are pregnant or potentially pregnant should avoid all alcohol consumption.

Cocaine (C) has been shown to be teratogenic in animal studies. Studies from pregnant humans are more complex, due to the difficulty of isolating cocaine abuse as a single etiologic factor. Cocaine-abusing pregnant women frequently have other substance-abuse problems or health problems that place their fetuses at risk. Some studies suggest that there may be an increased incidence of craniofacial, CNS, cardiovascular, and genitourinary anomalies. Cocaine use can lead to vasoconstriction in placental blood vessels, increasing risk for placental abruption.

Heroin (D) has not been firmly established as a teratogenic drug. Pregnant heroin addicts often expose their fetuses to other known teratogens. Since heroin readily crosses the placenta, newborn infants of heroin-addicted mothers will also be addicted and will exhibit opiate-withdrawal symptoms.

Rubella virus (E) is the classic example of a teratogenic infectious agent. In mothers infected with rubella virus during the first three months of gestation, approximately 20% of their fetuses become infected. Rubella virus crosses the placenta and infects the fetus, causing deafness, congenital cataract, and heart defects.

Herpes simplex virus (F) is not a potent teratogen, although some studies suggest that it may cause a variety of CNS abnormalities. In pregnant women actively shedding herpes simplex virus from vulvar lesions, the greatest risk is transmission of the disease to the newborn during vaginal delivery. Herpes simplex infections in newborns are often fatal, due to significant immaturity of the T cell system. Cesarean delivery is indicated for such mothers.

Toxoplasma gondii (G) is an intracellular parasitic protozoan that is known to be teratogenic. Maternal infection occurs by ingesting poorly cooked infected meat or exposure to animal feces. *Toxoplasma* crosses the placenta, infecting the fetus and causing chorioretinitis, microcephaly or hydrocephaly, and calcifications around the brain ventricles (often visible on ultrasound). Mental retardation and microphthalmia are also found in fetal toxoplasmosis.

Accidental exposure of pregnant women to poisoned fish in Japan and fungicide-contaminated grain in Iran established that **organic mercurial compounds** (H) are potent teratogens, causing abnormal development of the cerebrum and cerebellum as well as developmental defects in the peripheral nervous system.

Synthetic progestins (I) are not potent teratogens, but there is some evidence that suggests that progestins can cause masculinization of the external genitalia of female fetuses.

High doses of **ionizing radiation** (J) were once used on pregnant women. Radiation treatments for cervical cancer during pregnancy resulted in a high incidence of CNS abnormalities in fetuses. Similarly, among those surviving pregnant Japanese women exposed to atomic bomb blasts during World War II, there was a high incidence of spontaneous abortion and mental retardation in those children that were delivered alive. Even the small doses of radiation received with diagnostic X-rays or radioisotopes should be minimized in pregnant women. **Table 10.1** provides a convenient summary of other potentially teratogenic prescription medications.

TABLE 10.1
CLINICAL FEATURES OF SOME IMPORTANT
TERATOGENIC PRESCRIPTION DRUGS

Drug	Possible Teratogenic Effects
diethylstilbestrol	Causes vaginal and cervical CA in women exposed *in utero*
tetracycline	Tooth defects
streptomycin	Deafness
warfarin	Cartilage and CNS defects
trimethadione	Growth retardation, facial clefts, cardiovascular, genitourinary, and limb defects
phenytoin	Growth retardation, CNS defects, mental retardation
valproic acid	Craniofacial, cardiac, and musculoskeletal defects
aminopterin, also methotrexate	CNS and skeletal defects
tumor inhibiting drugs	Potent teratogens
ACE Inhibitors	Oligohydramnios and renal abnormalities
retinoic acid	Facial clefts, micrognathia, heart and neural tube defects
benzodiazepine	Fetal dependence, craniofacial defects
lithium	Cardiovascular defects

Items 553-559

Items 553-559

Match each of the mechanistic descriptions in the items below with the specific congenital anomaly that it causes. Answers may be used once, more than once, or not at all.

(A)   Phenylketonuria
(B)   Sickle cell anemia
(C) β-thalassemia
(D)   Down syndrome
(E)   Neural tube defect
(F)   Mandibulofacial dysostosis (Treacher Collins syndrome)
(G)   Cri du chat syndrome

553.   Chromatid deletion

554.   Autosomal translocation

555.   Multifactorial inheritance

556.   Point mutation in hemoglobin structural gene

557.   Autosomal dominant mutation, variable penetrance

558.   Decreased phenylalanine hydroxylase activity

559.   Mutation in globin promoter region

ANSWERS AND TUTORIAL ON ITEMS 553-559

The answers are: **553-G; 554-D; 555-E; 556-B; 557-F; 558-A; 559-C**. **Phenylketonuria** (A) occurs in 1/15,000 live-births in the US. It is caused by an autosomal recessive mutation and results in decreased hepatic phenylalanine hydroxylase activity. Routine screening at birth can detect this mutation and placing affected children on a phenylalanine-restricted diet can prevent the mental retardation caused by elevation in blood phenylalanine and subsequent decrease in CNS myelination.

Sickle cell anemia (B) was one of the first congenital human diseases to become understood on the molecular level. It is caused by a missense mutation which changes the glutamate codon to the valine codon in the sixth residue of the ß chain of hemoglobin to form hemoglobin S (Hb S). This substitution of a hydrophobic for a hydrophilic amino acid renders hemoglobin less soluble, especially at low oxygen tension. Precipitated hemoglobin forms large crystalline deposits and secondarily generates the crescent-shaped erythrocytes found in the blood smear. These abnormal erythrocytes are destroyed more readily in the spleen, leading to hemolytic anemia. The formation of clots leads to infarcts and these cause damage to many organs such as the spleen. Compromise of splenic function leads to recurrent infections. Vascular damage is also commonly observed in the joints, retina, skin and liver.

β-thalassemias (C) are diseases caused by abnormal synthesis of the ß-chains of hemoglobin. Genetic mutations that induce thalassemias represent a spectrum of abnormalities in globin genes including transcription, RNA splicing, mRNA stability and translation. Over 50 different mutations have been reported in the ß-globin gene. ß°-thalassemia is characterized by a complete absence of ß-globin caused by either single base mutations that give rise to translation termination or short deletions and insertions in the ß-globin coding sequence altering the reading frame. ß$^+$-thalassemia is characterized by reduced levels of synthesis of ß-globin chains. Some ß$^+$-thalassemias are due to mutations within the ß-globin promoter region which result in an 80% reduction in the levels of ß-globin mRNA.

Other ß-thalassemias are caused by mutations that lead to abnormal splicing of the globin mRNA. Mutations involving single base substitutions in the critical 5′ or 3′ splice junctions lead to cryptic splice sites resulting in incorrectly spliced mRNA that is either untranslatable or is translated into nonfunctional proteins. Aberrantly spliced globin mRNA can be detected by S1-nuclease protection assays or Northern blot analysis of the mutant mRNA.

Down syndrome (D) is caused by trisomy of chromosome 21. Most cases are caused by primary nondisjunctional trisomy but about 5% of the cases are caused by translocation of chromosome 21 onto another chromosome, typically 14. Translocation Down syndrome and the trisomic version are phenotypically similar.

Neural tube defects (E) include a complex collection of disease entities characterized by failure of closure of the neural tube and the associated bony structures of the skull and vertebral column. If there is extreme cranial closure failure and lack of formation of the vault of the skull, an anencephalic fetus will result. Less severe neural tube defects such as spina bifida occulta involve only abnormal formation of the vertebral arches in the lumbar region and may be asymptomatic. There is both a genetic and an environmental etiology to neural tube defects. Thus, their incidence follows a multifactorial inheritance pattern.

Mandibulofacial dysostosis (Treacher Collins syndrome) (F) is most often caused by an autosomal dominant mutation of variable penetrance. The development of the first pharyngeal arch is most affected so that afflicted patients have microtia, midfacial hypoplasia, downward sloping palpebral fissures, lower eyelid coloboma, and conductive hearing loss. Patients are sometimes mistakenly thought to have mental retardation but slowness to develop normal speech is due to hearing difficulties.

Cri du chat syndrome (G) is an uncommon genetic entity characterized by microcephaly, profound mental retardation, and a cat-like cry during infancy. Most cases are caused by a deletion of the short arm of chromosome 5 (monosomy 5P).

Choose the **BEST** response.

560. All of the following drugs have been established as clear teratogens **EXCEPT**:

    (A)   alcohol
    (B)   heroin
    (C)   diazepam
    (D)   lithium
    (E)   synthetic progestins

561. Which dose of alcohol is safe for consumption by a pregnant woman?

    (A)   1 drink/day
    (B)   2 drinks/day
    (C)   4 drinks/day
    (D)   widely separated binges
    (E)   abstinence

562. The developmental process **MOST** likely affected by excessive alcohol consumption during pregnancy is

    (A)   neural tube closure
    (B)   neural crest cell migration
    (C)   heart septation
    (D)   fusion of pleuroperitoneal membrane
    (E)   apoptosis

ANSWERS AND TUTORIAL FOR ITEMS 560-562

The answers are: **560-B; 561-E; 562-B**. All of the drugs listed in Item 560 above are known teratogens except for **heroin**. Although heroin has not been clearly established as a teratogen, many pregnant heroin addicts also consume alcohol, nicotine, cocaine, and other drugs. Maternal nutrition may also be poor, leading to a poor developmental outcome for the fetus. The child of a heroin-addicted mother will be addicted as well at birth because heroin and other opiates readily cross the placenta. Heroin-addicted newborns will also experience withdrawal symptoms.

    Alcohol consumption during pregnancy is **contraindicated**. No dose is known to be safe so current guidelines recommend alcohol abstinence during pregnancy. Fetal alcohol syndrome (FAS) is seen in children of alcohol-abusing mothers. Children with FAS have a plethora of problems including CNS abnormalities such as microcephaly and mental retardation, intrauterine growth restriction and small stature, and a complex of minor midfacial hypoplasia. Some experimental studies in animal models suggest that alcohol or its metabolites alter cell migration during development. Altered neural crest cell migration may account for the midfacial hypoplasia commonly seen in FAS.

Items 563-565

A 34-year-old woman who is a severe, chronic alcoholic gave birth five years ago to the child shown in **Figure 10.6**. The child exhibits a pattern of defects associated with fetal alcohol syndrome (FAS).

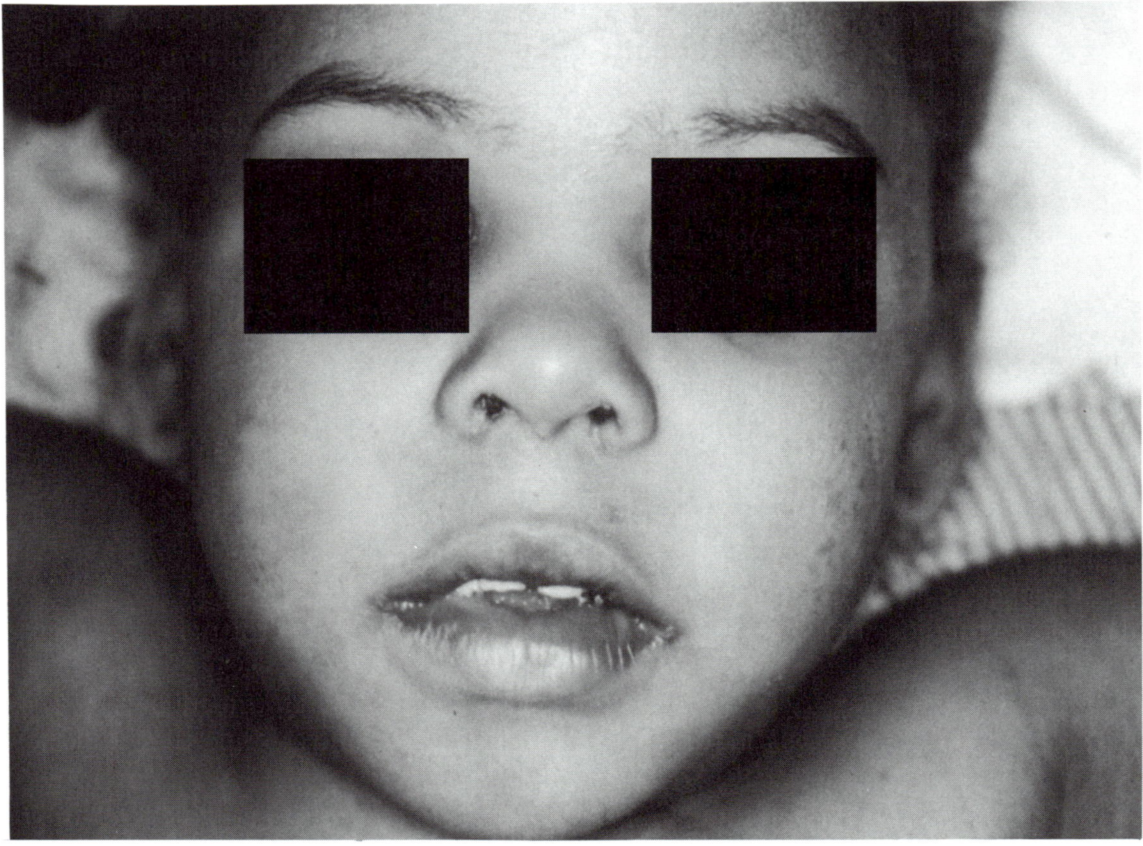

Figure 10.6

563. All of the following facial characteristics are usually associated with this syndrome **EXCEPT**:

(A) maxillary hypoplasia
(B) long palpebral fissures
(C) thin upper lip
(D) epicanthal folds
(E) short nose

564. Which of the following statements concerning the fetal effect of maternal alcohol consumption during pregnancy is **TRUE**?

(A) Moderate maternal alcohol consumption has no effect.
(B) Maternal alcohol abuse is thought to be the most common cause of mental retardation.
(C) Heavy alcohol consumption limited to a few days during pregnancy is thought to have no effect.
(D) The susceptible period of brain development is limited to the first trimester.
(E) Fetal alcohol effects (FAE) usually do not involve brain functions.

565. Which basic developmental mechanism **MOST** likely accounts for the craniofacial anomalies seen with fetal alcohol syndrome?

(A) abnormal hox gene expression
(B) abnormal neural crest cell migration
(C) abnormal somatic segmentation
(D) abnormal induction of central nervous system
(E) abnormal apoptosis

ANSWERS AND TUTORIAL ON ITEMS 563-565

The answers are: **563-B; 564-B; 565-B**. A child afflicted with a severe case of **fetal alcohol syndrome** (FAS) is shown in **Figure 10.6**. This child exhibits a specific pattern of craniofacial defects. The craniofacial features of FAS include short rather than long palpebral fissures although the palpebral fissures of the child shown in Figure 10.6 are not affected. However, this child's face does exhibit **maxillary hypoplasia**, thin upper lips, epicanthal folds, and a short nose. Abnormal palmar skin lines and **mental retardation** were also present. Other associated defects are microcephaly, prenatal and postnatal growth deficiency, and heart and joint anomalies.

The incidence of FAS is estimated to be 2 per 1000 live births and is considered to be the most common cause of mental retardation. Moderate maternal alcohol consumption may

213

produce **fetal alcohol effects** (FAE) and may damage the developing brain, resulting in infants with behavioral and learning disabilities. Binge drinking will produce FAE.

Research on laboratory animals suggests that the root cause of FAS is abnormal **migration of the neural crest cells** that populate the pharyngeal apparatus. In addition, the CNS effects of FAS may also be due to disruptions in cell migration. During development of the cerebrum and cerebellum, there is a great deal of migration of neuroblasts. Disruption of this process may result in mental retardation and other behavioral effects such as attention deficit disorder and hyperactivity, commonly observed in children with FAS

<u>Items 566-569</u>

Choose the **BEST** response.

566.   The period during pregnancy **MOST** sensitive to teratogens is

      (A)   preimplantation
      (B)   first trimester
      (C)   second trimester
      (D)   third trimester
      (E)   fetal

567.   All of the following are likely a cause of spontaneous abortion **EXCEPT**:

      (A)   chronic illness of the mother
      (B)   fetal genetic defects
      (C)   maternal uterine anomalies
      (D)   intercourse in early pregnancy
      (E)   progesterone deficiency

568.   All of the following chemicals are known to be a human teratogen **EXCEPT**:

      (A)   lead
      (B)   polychlorinated biphenyls (PCBs)
      (C)   methyl mercury
      (D)   lithium carbonate
      (E)   caffeine

569.   At what week of gestation is exposure to rubella (German measles) likely to do the **MOST** harm to the conceptus?

    (A)   6 weeks
    (B)   14 weeks
    (C)   20 weeks
    (D)   28 weeks
    (E)   32 weeks

ANSWERS AND TUTORIAL ON ITEMS 566-569

The answers are: **566-B; 567-D; 568-E; 569-A**. The **first trimester** is the most sensitive period because it is the time when organogenesis occurs and when cell division and differentiation are at their peak. During this period teratogens can cause major malformations. They do not cause defects during the two weeks after fertilization but teratogens that interfere with cleavage of the zygote and implantation of the blastocyst cause death and spontaneous abortion of the conceptus.

    Intercourse during early pregnancy is not considered a cause of spontaneous abortion. All of the other choices have been shown to cause spontaneous abortion. All of the chemical listed in Item 568 except **caffeine** are know to be teratogenic in man. Caffeine is the most widely consumed stimulant and is present in many beverages. Heavy consumption during pregnancy may be unsafe and should be avoided. The earlier in pregnancy the maternal **rubella** infection occurs, the greater the chances that the embryo will be malformed.

INDEX OF CLINICALLY IMPORTANT TOPICS AND MAJOR DEVELOPMENTAL CONTROL MECHANISMS

E-G

H-J

hernia, congenital diaphragmatic, 25, 102
        , physiological, of midgut, 130
hydrocephalus, 47
hydrometrocolpos, 161
hygroma, cystic, 205
hyperplasia, congenital adrenal, 164, 171
hymen, unruptured, 161
hypertrophy, right ventricular, 124
hypoplasia, thymic, 66
        , pulmonary, 102
hypospadias, 169
infections, *Pneumocystis*, 65
isoimmunization, Rh, 67
jaundice, 68
        , obstructive, 127
jelly, Wharton's, 186

K-N

karyotype, normal, 200
kidney, congenital polycystic, 155
        , horseshoe, 151
malformation, Arnold-Chiari, 47
mandibulofacial dysostosis, 77, 85, 210
meconium, 129, 138
mercury, organic, 207
mesoderm, derivatives of, 16
microphthalmos, 93
midgut, rotation, 133
mole, hydatidiform, 181
movement, fetal, 57
myelomeningocele, 47
nucleus pulposus, 25
neural crest, 43
        , derivatives of, 16
        , fetal alcohol syndrome, 214
        , in cardiac development, 106
nondisjunction, 192, 194, 197, 204

O-R

oogenesis, 5
ossification, endochondral, 21, 23
        , intramembranous, 23
osteogenesis imperfecta, 30

S

RAPID PREPARATION

for the

USMLE
Step 1

FINAL REVIEW
for the Medical Boards

Featuring:

- Top-grade pictures like those seen on the USMLE Step 1
- On-target coverage of the highest-yield topics tested
- Clinical scenarios and problem-based questions
- Single-best and extended questions
- Detailed, annotated answers to all questions

Rapid Preparation for the USMLE Step 1

The Princeton Review, the nation's leader in test preparation has teamed up with J&S Publishing to provide medical students and graduates with an intense program of study during the final weeks before the actual exam. Rapid Preparation for the USMLE Step 1 covers the topics most often tested on the exam. By completing each of the questions in this book and reviewing the tutorials, you'll be in top shape, ready to handle the integrated, clinical scenarios that pervade the exam.

Our innovative and expansive book also features the most comprehensive presentation of the MRIs, electron micrographs, and other high-quality printed images often seen in the second and fourth booklets of the USMLE Step 1.

If you want to get an early start, you should also consider purchasing the entire catalog of books published by J&S. Each book covers one of the basic sciences and contains more than 500 practice items prepared like the USMLE Step 1.

The Princeton Review offers live preparatory courses for the USMLE Step 1, featuring the no-nonsense personal attention featured in courses, books, and software, that has helped millions of students and graduates attain their academic and professional goals.

Whether you're preparing for the Step 1 for the first time or you've been out of medical school for an extended period of time, *Rapid Preparation for the USMLE Step 1* will help you achieve a great boards score.

Also published by J&S Publishing Company, Inc.:
Anatomy: Review for the New National Boards • Microbiology: Review for the New National Boards • Biochemistry: Review for the New National Boards • Pathology: Review for the New National Boards • Cell Biology: Review for the New National Boards • Physiology: Review for the New National Boards • Pharmacology: Review for the New National Boards • Surgery: Review for the New National Boards • The Study of Surgery: Comprehensive Clinical Tutorials in Problem-Solving for Anatomic and Specialty fields of General Surgery • Obstetrics and Gynecology: Review for the New National Boards

J&S Publishing Company, Inc.
1300 Bishop Lane
Alexandria, Virginia 22302
phone: (703) 823-9833
fax: (703) 823-9834
www.jandspub.com
jandspub@ix.netcom.com

ISBN 1-888308-02-8

For USMLE course information, contact
The Princeton Review
1-800-2REVIEW
www.review.com
info.tpr@review.com